THEORY SAYS

WHO
OWNS OUR
UNIVERSE

by
Trillion Theory author
Ed Lukowich

Cosmology book series (non-fiction) by Ed Lukowich:
Trillion Years Universe Theory, (cosmology) 2014
(Republished December 2015).
Trillion Theory (cosmology) October 2015.
Black Holes Built Our Cosmos (cosmology) December 2015.
T Theory Says: Who Owns Our Universe (cosmology) Nov 2016.

Science Fiction novel authored by Ed Lukowich:
The Trillionist, 2013 under pen name Sagan Jeffries.
Publisher: Edge Science Fiction & Fantasy Publishing.

Library and Archives Canada Cataloguing in Publication
Lukowich, Ed 1946 – T Theory Says: Who Owns Our Universe
ISBN: 978-0-9918408-8-5 (paperback) (Trillion Theory Series)
ISBN: 978-0-9918408-9-2 (e-book) (Trillion Theory Series)
1. Cosmology. I. Title. QB981.L823 2016 523.1
C2016-906598-7 C2016-907173-1 Printed by CreateSpace.

Founder of T Theory

Canadian theorist Ed Lukowich (Trillionist), presents a new theory, uncovering major truths about our cosmos. Years of study led to writing T Theory (TT), which depicts our cosmos to be a trillion years of age. Ed admits, *'My theory is a long shot, new and very different. Yet, my goal is to spread the word about my new TT universe theory and convince the skeptics.'*

TT states, *'A trillion years ago, our cosmos was scientifically designed to perpetually recycle its stars and solar systems as a way of growing its galaxies ever larger. While our cosmos began small, it is now mighty, having grown to 73 quintillion stars. The new concepts in TT endeavor to bring an end to acceptance of the Big Bang paradigm, and all the incorrect notions that our cosmos is only 13.7 billion years old.'*

Mark Twain said, *"The two most important days in your life are the day you were born and the day you find out why."*

Ed says, *'I was born to reveal the truths and shed new light as to how, when, and why our cosmos came to be. My Trillion Theory displays the wondrous cosmos as a grand scientific enterprise, invented by incredible scientific genius.'*

Besides his four cosmology books, Ed also authored a sci-fi novel entitled **The Trillionist**, under his pen name Sagan Jeffries. The novel was published in 2013 by EDGE Science Fiction & Fantasy Publishing. That futuristic novel depicts the actions of a self-prevaricating entity older than dirt, with a history going back a trillion cosmic years.

See www.trillionist.com

T Theory Says is the 4th in Ed's cosmology series: *Trillion Years Universe Theory,* 2014; *Trillion Theory,* 2015; *Black Holes Built our Cosmos,* 2015; *T Theory Says Who Owns our Universe,* 2016;
All available at Amazon, in paperback and ebook.

New ideas that Trillion Theory lays claim to:
- depicts an ancient trillion-year old cosmos.
- shows a black hole residing in Earth's core.
- an XL black hole at the center of our Sun.
- a black hole at the core of all cosmic spheres.
- a black hole provides spin to its sphere.
- a black hole provides a sphere with its gravity.
- black holes built the spheres they occupy.
- black holes recycle spheres and solar systems.
- on the outer perimeters of cosmos, there exists a vast ocean of energy which black holes can access to continually build our cosmos larger.

If you enjoy new cosmology ideas, you'll relish this opportunity to read this new theory of how our cosmos originated and operates.

TT promises to surprise you by digging deep, uncovering more of the cosmic secrets about, 'Who Owns our Universe?'

Expect these surprises in T Theory.

TRILLION YEARS
Expect a universe a trillion years old.

CYCLES, each about 15 BILLION YEARS.
Expect 67 cycles in a trillion years.

A BIGGER COSMOS IN EACH CYCLE.
Expect an ever-growing cosmos.

COSMIC SPHERES WERE BUILT.
Expect that black holes are the machines
which built all of the cosmic spheres.

A BLACK HOLE RESIDES INSIDE EARTH.
Expect a black hole inside of each
and every spheroid across all our cosmos.

SPIN AND GRAVITY.
Expect a black hole to provide the spin
and gravity inherent to each cosmic sphere.

TT outlines its major issues to be answered:

1. How did black holes build the spheres of our cosmos, and now control all of the spheres, solar systems and galaxies within our cosmos?

2. How are black holes artisaned to be builders? **TT says,** 'Black holes are cosmic engines that eat light, spinning it into matter, in order fashion an orb of matter around themselves.' Thus, TT shows a hidden (cloaked) black hole inside every sphere, including Earth. **TT says,** 'Solving secrets of black holes is vital in climbing high enough to peer over the fence into the artisan's backyard.'

3. What were the other main strategic inventions in the owner's master plan in building a cosmos? **Light,** with its array, is the second great invention.

4. Who owns these inventions and our universe? Careful not to trample past the obvious. **TT says,** 'Never rule out the possibility that a so-called Black Hole Society comprised of trillions of black holes throughout our cosmos, is an actual owner of our universe. Or, if black holes are just cosmic builders, then what genius scientifically designed and manufactured these incredible machines?'

Definitely, building a cosmos wasn't done in a day. It has taken a trillion years of cosmic history.

Quest of T Theory is to uncover universe secrets. The end goal: Who built and owns our universe?

Not like presto, creation; rather as a builder, inventor, designer, engineer, architect, deploying super science. Within TT, is evidence of incredible science being used throughout the cosmos. TT claims that the design of our cosmos came from the mind of a very clever savvy shrewd strategist.

T says, *'Ever since cosmos began a trillion years ago, its materials and engines perform to recycle solar systems (and all their spheres) within ever-growing galaxies. This recycling is so cleverly set, that since the cosmic inception, the owner has never had to deploy extra manipulation, needing to run out to turn a crank, or re-charge a battery.'*

Thus, TT goes where no man has gone before - in search of a cosmic owner. TT is armed with the mayor questions: How? and Why?

Here is one possibility? Suppose outside of this universe is a timeless existence which ebbs on to eternity. So, an adventurous owner decided to construct our cosmic oasis where entities come to experience beginnings and ends, done within the nervous tense experience we know as *time.*

TT shows how *time* was put into our universe.

T Theory trumps the Big Bang.

Criteria needed for a correct universe theory:

- The 1st criterion is that a correct universe theory must be totally explainable; whereas the Big Bang fails to explain a profusion of cosmic phenomena.

- The 2nd criterion is that a correct theory must encompass all, a Theory of Everything, leaving no stone unturned; whereas Big Bang's inadequacies leave a plethora of unsolved queries. Yet, people are blindly enthralled by the Big Bang.

Carl Sagan said, 'It is better to grasp the universe as it really is, than to continually persist in delusion, however satisfying and reassuring.'

- The 3rd criterion is that a correct theory must be totally provable, beyond any shadow of a doubt; whereas the present supposed proofs for Big Bang are not proofs at all, failing all critical acid tests. Rather, they are twisted findings which simply aspire to aid an unprovable Big Bang.

T Theory will show how Big Bang over-shot the runway, being a hoax which has sucked in millions of people, and sadly slowed cosmic discovery for over a half century. TT looks to rectify this. TT reaches out to speed our cosmic progress.

Next, are some of TT's introductory basics which boldly add into what we already know of cosmos.

T Theory (TT) states its BASICS

T

- A black hole eats light spinning it into matter.
- A black hole spins light into a body of matter.
- A black hole (BH) is a cosmic sphere builder.
- A cloaked BH resides inside of every sphere.
- BH spheres are planets, moons, and suns.
- BH resides billions of years inside a sphere.
- A residing cycle can last up to 15 billion years.
- A BH sheds its sphere by its cycle's end.
- A black hole is totally indestructible.
- A sun's black hole can survive a Supernova.
- BHs survive Obliteration of their solar system.
- A BH always survives the death of its sphere.
- A BH splits (replicates) at its survival moment.
- An exposed BH always rebuilds a new sphere.
- A black hole can build 1 sphere in each cycle.
- A black hole can build a sphere in every cycle.
- Black holes populate their solar systems.
- Solar systems recycle within their galaxies.
- Solar system numbers increase inside galaxies.
- Ancient galaxies grow in numbers and in size.
- Recycling was (is) on-going for a trillion years.

Origin of our Cosmos

TT says, 'At the cosmic origin, all that existed was a vast frozen non-moving endless ocean of light (space didn't yet exist). Then, interjected into the ocean was a naked hungry black hole. To free some light for its eating, the spinning black hole broke chunks from the ocean wall, spinning and capturing light into atoms of matter round itself.'

In TT, we chase the path of prolific growth from one black hole building the universe's first orb to the trillions of black holes now in control of all cosmic spheres - an enormous evolvement which took place over the past trillion cosmic years.

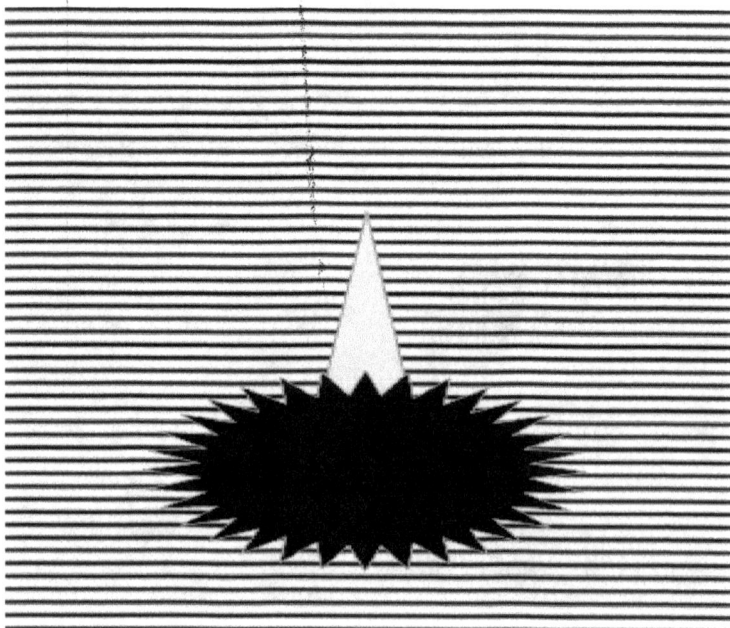

T Theory maps a cosmic time progression. Cosmos began from 1 black hole eating light. Then over time, millions of black holes got newly introduced into the ocean of light.

Cosmic origin (Cycle 1), a black hole ate ocean light spinning it to matter around itself.

First 105 billion years (Cycles 1-7) more black holes spun light into spheres in solar systems.

From 105-300 billion years, the largest solar systems recycled to form small galaxies.

From 300-600 billion years, spheres and solar systems recycled to form larger galaxies.

From 600-900 billion yrs, cosmos exponentially grew giant galaxies in an expanded space.

In year 1 trillion, in cycle 67, our present 15 billion year cycle houses 73 quintillion stars inside of billions of humongous galaxies. Solar systems populate these ancient galaxies throughout the cosmos.

This circumference represents a single15 billion-year cycle as the upper time limit during which a black hole maintains control of the sphere it spun around itself. Then, next cycle, it builds a new orb in a new solar system.

TT shows a cosmos which is a trillion years old, where the recycling of spheres and solar systems is paramount to the continual growth of galaxies.

Astronomers tell us there are millions to billions of these galaxies. TT says, 'Each gigantic galaxy is controlled at its hub by a SUPERMASSIVE black hole. Each supermassive controls millions of solar systems within the galaxy. While a supermassive is 200-800 billion years old, solar systems inside of a galaxy all live by a 15 billion year upper limit within which they die and then recycle into new spheres in new more numerous solar systems.'

T Theory (TT) cycle says:
- Cosmos is a trillion years old.
- Spheres can only last at most 15 billion years.
- Cosmos has had 66 such cycles. We are 67.
- Sometimes, solar system spheres do die and recycle earlier at 10-13 billion years.
- A sun going Supernova prematurely shortens the 15 billion year cycle of its solar system.
- At the end of a cycle for a solar system, the black holes inside of the spheres survive.
- The survival black holes always start up a new solar system for the next 15 billion year cycle.

T Theory tells the intriguing story of how even the very oldest known stars make our cosmos appear to be a youngish 13.7 billion years, yet the supermassive black hole at a galaxy's hub hides its actual 800 billion to trillion year age.

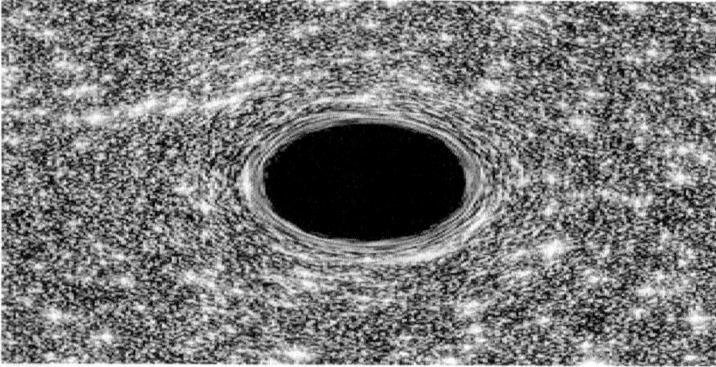

T Theory says, 'Examine a colossal galaxy. All the suns and solar systems orbiting the supermassive black hole are younger than 15 billion years. Yet, the supermassive is 500-800 billion years old.'

But, please don't get too hung up on exactness of 15 billion years, or 67 cycles, or a trillion years. More important are the repetitive ideas within T Theory, such as: light spinning to form matter; a black hole living inside of every sphere; orbs and solar systems recycling within ancient galaxies.

T Theory says, 'Galaxies and our entire cosmos, grow exponentially on an approx 15 billion-year recycle plan. Each new cycle generates many more new suns and solar systems within ancient galaxies. TT maintains that there are trillions of alive black holes providing cosmic recycling.'

T Theory says, Think of our cosmos as a living entity, recycling and growing at a snails pace over

Origin, a trillion years ago, all that existed was an endless ocean of frozen light.

Just the current rendition (generation) of stars and solar systems inside of more aged galaxies. Even today, in a present 67th 15 billion-year generation, in the trillionth cosmic year, millions of stars will go supernova and destroy their solar systems. Yet, the black holes of a destroyed solar system can never be destroyed. Instead, they will reform a brand new larger solar system to replace the old.'

Whoa you say, 'Growth by our cosmos requires extra matter. But, new matter cannot be created, as the amount of matter available to our cosmos was supposedly set at the moment that a Big Bang exploded all matter into waiting space.'

TT says, That assumption is wrong.
Quit thinking in Big Bang terms.
For, TT shows that cosmos has an inexhaustible supply of energy, accessible by black holes.
So that black holes can build more spheres inside of more solar systems inside of growing galaxies.'

Follow a couple of steps in grasping T Theory.

During a trillion years, billions of black holes ate from the light ocean, freeing tons of light.

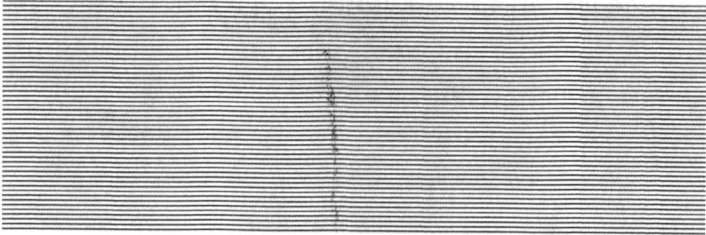

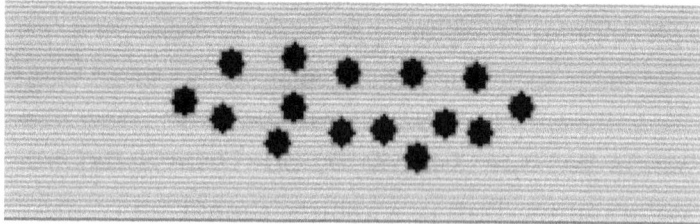

The black holes tightly spun that light from the ocean into compressed spheres around themselves. A vast cavern of empty space resulted inside of the frozen light ocean.

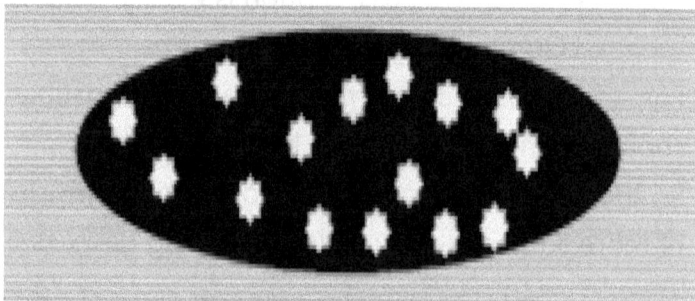

The **GOALS of T Theory** are genuine:
- Provide a more succinct and accurate theory to explain the origin and functions of our cosmos.
- Disprove other theories such as Big Bang and Nebular which have been wrongly accepted.
- Show proofs for TT which can be validated in the future by astronomers and astrophysicists.

Some PROOFS for TT are hidden amongst the suns, planets, moons, solar systems and galaxies. TT says, 'Black holes hold the precious key.'

T Theory attacks the Big Bang theory
Now the question is, can T Theory shed enough proof about our cosmic origin (age) to dethrone the Big Bang? TT says, 'It's time to re-examine how we see our cosmos. Is Big Bang near adequate?' A Diane Grant quote, "It's better to walk alone, then with a crowd going in a wrong direction."

TT denounces the absurd Big Bang which fails to explain how solar systems and galaxies formed; or where atoms came from, how they were spun, and how they continue to spin for eons of time.

TT says, 'Big Bang is archaic; a sacred cow ready to be tipped over and disproven. Big Bang is a dud when it comes to explaining the cosmos.

T Theory says, It is imperative for astrophysicists and for astronomers to agree that the Big Bang theory should no longer be the paradigm model.'

No to Big Bang

Many prominent people are in agreement. 'Astronomers now admit they must re-think our universe. Recently, they discovered numerous large cracks in the Big Bang theory.' Origin via an explosion fails to explain the workmanship that went into cosmic complexities. Astronomers are wrong, estimating cosmos to be only 13.7 billion years old. As well, Big Bang and Nebular theory are wrong as to how our cosmos formed.

Whereas, T Theory affirms that our cosmos has grown, escalating in size and sphere population with each new 15 billion-year next-generation of planets, stars, and solar systems within galaxies. T Theory estimates 67 of these 15 billion recycles occurred over a trillion years of cosmic history.

In 2004, a Gemini Telescope found galaxies more fully mature than one might expect in a cosmos of only 13.7 billion years. **Dr. Robert Abraham, Dept of Astronomy, U. of T.** 'We are seeing that a large fraction of stars in cosmos were in place when the universe was young, that should not be the case. This glimpse back in time shows pretty clearly that we need to re-think what happened.'

Dr. Patrick McCarthy, Carnegie Inst. added, 'It's unclear if we need to tweak the existing models or develop a new one to understand this finding.'

Whereas, T Theory answers these mysteries. TT works diligently to earn support as being the best new explanation of our cosmos. More and more, these discoveries are pointing to T Theory.

T Theory says, 'Brand new theory vastly expands our minds. No great idea was ever immediately accepted; most are initially viewed as ridiculous.'

But, someone stipulates that there are proofs for Big Bang. For the past 70 years, astrophysicists have tried to find proofs for the Big Bang such as: redshift readings indicate that galaxies ebb away from each other; microwave background radiation indicates an explosion glow; gravitational waves mean expansion. All unacceptable to T Theory.

X-out and say NO to Big Bang.

Big Bang's 13.7 billion year age estimate for cosmos was jeopardized in 2014. Astronomers discovered the 15.5 billion year old Methuselah Star in our Milky Way. How could a single star out-age the cosmos? They couldn't explain such a paradox. A fly in Big Bang's ointment.

Do you know who named the Big Bang?
In 1949, English astronomer Fred Hoyle in a BBC radio broadcast rejected the idea of an explosion theory as cartoon physics, mockingly coining it Big Bang, calling the theory a ridiculous hoax.

TT says, 'Keep an open mind. Past history shows misconceptions leading people to believe that Earth was the cosmic center, or that our Earth was flat - falsehoods that prevailed for centuries.

TT predicts that the same fate awaits Big Bang after new proofs support a correct theory.'

TT says, 'No, to the Big Bang hoax which blows smoke while dragging scientists around for far too long. Bang's a poor effort in a modernistic world. Bang is a fallacy, built upon pillars of sand. When wiser future generations look back, they will laugh at the Bangers, just as we did towards those who believed in a flat Earth.'

T Theory says false to the Big Bang theory for:
  • estimating the cosmic age at only 13.7 billion years by incorrectly using just the current stars.
  • depicting cosmic origin as coming from an explosion of matter expanding outwards.
  • concluding that galaxies receding from each other proved an explosive origin to our cosmos.
  • accepting a Nebular theory which incorrectly proposes that solar systems/galaxies are the result of nebular clouds swirling and cooling.

TT says, 'It's foolhardy to place a tag such as Big Bang onto a scientific invention as intricate and awesome as cosmos, where wondrous similarities are found in all the solar systems and galaxies. To accept an idea that an iota of matter somehow exploded, bursting cosmos into existence, is a ridiculous stretch. Furthermore, galaxies receding away from each other is insufficient Bang proof.

TT shows that there are other more practical reasons why our cosmos is expanding outwards.'

Did you know that the renowned pioneer Radio Astronomer Grote Reber had always been highly skeptical, stating that Big Bang was bunk.

TT states, 'A fallacy Big Bang had it all wrong, with no way to show how spheres, solar systems, and galaxies formed. So, Bang then recruited Nebular Theory to try to show how 4.6 billion years ago hot gas clouds supposedly swirled, collapsing and vortexing into cooled spheroids to form our solar system – and others.'

TT says, 'Nebular Theory is total bunk. Cosmos has workmanship beyond pure chaotic chance. Till T Theory, no one has been close to figuring out cosmos, so they accepted the excitement of a Big Bang explosion and Nebular swirling.'

Nebular asks us to believe that spin swirled hot matter into atomized spheres. T Theory says, 'One solar system is pure chance, but billions is a designed scientific methodology.'

Like dummies we're fed convenient answers by astrophysicists which fit nicely into Big Bang and Nebular Theory. But, how come suns stay hot, not succumbing to the coldness of space, while planets and moons, because of a smaller size, supposedly went from being hot entities to cold surfaces? How come planets and moons aren't then frozen right to their core while sitting in cold space? How come Earth is hot in its center? How come most spheres have volcanic activity?

It's said, 'Compression of the surface started a fire deep in the pit of Earth.' T Theory won't buy that. Instead, T Theory tells the real reason why a fire started deep inside the pit of our Earth.

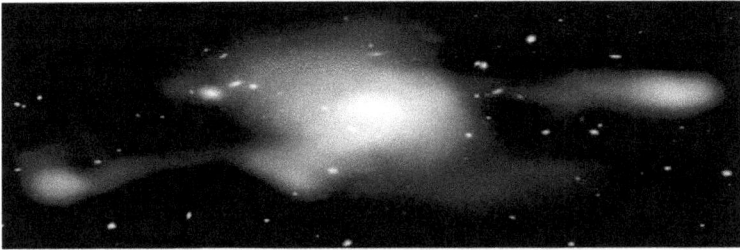

T Theory attacks Nebular Theory as well, saying, 'False to Nebular Theory assumptions as to how the planets, moons, and suns supposedly formed from the nebulae of swirling hot gas clouds.' In Nebular Theory, supposedly gas clouds from Big Bang coalesced. Stephen Hawking states that stardust of swirling nebula clouds gravitated and assembled all the gas particles into spheres in our solar system. But, Nebular is as crazy as Bang.

TT says, 'The real question is: what internal force is really causing the interior of our planet to heat up while it sits in the extreme coldness of space. What started a fire deep in Earth's bowels?'
    The TT solution is both multiplex and shocking.

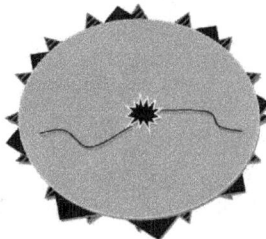

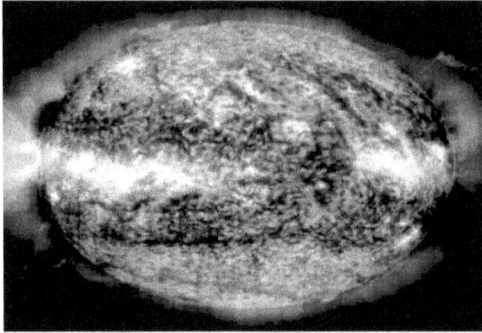

The TT answer, 'A difference between a sun and a hard spheroid (Earth) is that the sun's black hole has already lost the control game, as much of its spun matter is already unraveling back to light.'

 T Theory is the first to state, 'Light trapped as matter will always escape from the clutches of a black hole, even if it takes billions of years. The oldest matter of Earth, the matter at Earth's core, is unraveling in its attempt to escape from the black hole deep inside of Earth's core, then to reach the surface and escape back to being light. That is why volcanoes occur on planet Earth.'

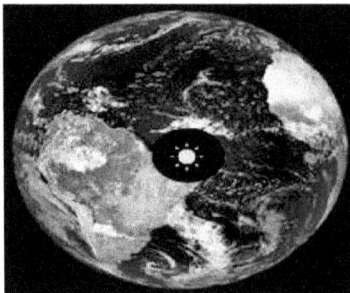

Time for a new universe theory
TT says, 'The time is ripe for the acceptance of a superior theory at unlocking cosmic secrets. Bang and Nebular theories are both hoaxes depending upon the roll of the dice and chance. Both fail to explain multitudes of cosmic complexities.

In TT, 'Important Cosmic Rules of Engagement, which via gravity, seemingly regulate interactions between naked black holes and also between the spheres which black holes have built in cosmos.'

Today, we search for a true better way to explain our entire universe. **A Theory of Everything (ToE)** should encompass a true explanation of the entire spectrum of a cosmos, right from planets, moons, suns, solar systems, galaxies, gravity, light, space, atoms, and even black holes all under one super umbrella. Thereafter, we can tackle the question of why we are in it? Plus, who owns our universe?

Theory of Everything
T o E

Now, T Theory takes on Stephen Hawking, who is a fantastic spokesman for cosmology. However, he does get a few ideas wrong.

Stephen Hawking, the renowned English theorist and cosmologist, in April of 2013 admitted to a rather large blunder. Till then, he'd thought that light swallowed up by any black hole was forever lost. He has now recanted, admitting radiation can escape. But, he still hasn't figured out, as T Theory has, that light can escape after billions of years in captivity inside of a black hole.

One law of TT is that light will always escape a black hole and return back to free traveling light, even if it takes billions of years. T further states that light trapped inside a black hole spins as atoms for eons upon eons, until the black hole tires, eventually losing total control of its matter.

T Theory introduces the main concepts that must be allocated into its Theory of Everything. Already, it is commonly known that atoms spin as matter comprising spheres, that spin on their axis within solar systems, which spin inside of giant galaxies.

T Theory is the first and the only theory to tie all these actions to one common factor, black holes.
T Theory demonstrates this commonality:
- light is trapped and spun into spinning atoms of matter for eons, by a black hole.
- spheres (planets, moons, suns) spin on their axis because cloaked inside of them is a resident black hole.
- the revolution of any solar system comes from the powerful gravity coming from the sun's resident x-large black hole.
- galaxies, and their host of solar systems, are held in a spiraling formation by spin and gravity coming from the supermassive black hole at the hub of the galaxy.

TT says, 'All cosmic organization is a result of the talents, actions and interactions of black holes.'
TT further asserts, 'Billions of solar systems are inside of billions of galaxies. Solar systems are rampant across all of the cosmos, such that a sun without a solar system is a cosmic rarity.'

How come solar systems and galaxies are cosmic units? Big Bang can't explain. But, T Theory can.

To explain, we travel to the closest spiral galaxy to our Milky Way, to Andromeda, also known as M31. At night, Andromeda can be seen by the naked eye from Earth, appearing like just a single star. Yet, Andromeda holds billion of stars.

TT declares that, 'Andromeda has a million solar systems. The reason why is that the supermassive black hole at Andromeda's hub is as much as 800 billion years old, having started the galaxy that long ago. This 800 billion years is far older than a 13.7 billion years which astronomers estimate.

All of the solar systems within Andromeda are kids compared to the age of the galaxy. Most of Andromeda's solar systems are only 1-13 billion years old, while a few old solar systems are 14-15 billion years. The age of any Andromeda 'rental' solar system depends upon its cycle phase.'

TT says, 'We see huge age differential inside of the Andromeda Galaxy. At the hub of the spiral galaxy is a supermassive black hole, some 800 billion years old. While, the bright stars clustered around the supermassive are 10-13 billion years.

Andromeda started small 800 billion years ago (not quite back to the T Theory cosmic origin). Andromeda has multiplied in size over time, now displaying billions of occupant solar systems. Its supermassive black hole, a permanent fixture at 800 billion years of age, powerfully controls all of the billions of smaller sized black holes at the cores of the suns, planets and moons inside of the galaxy. These smaller spheres live by the 15 billion year rule; meaning they die and recycle at some time prior to their 15 billionth year.

Andromeda's supermassive black hole evolved by adapting its structure which gained it a grand permanency at the controlling helm of its galaxy.

TT illustrates this evolution later in this book.

So, for T Theory the goals to accomplish are:

♦ Show Big Bang theory as unproven theory.

♦ Show how T (Trillion) Theory (TT) is by far the correct theory and demonstrate how it answers huge questions as to age, origin, and operations of our planet, solar system, galaxy, and cosmos.

♦ Provide possible proofs for T Theory, and suggest how astrophysicists and astronomers can assist by looking for these 4-5 proofs that show T Theory as the new cosmic discovery theory.

Before more presentation of T Theory, and prior to seeking proofs for T Theory, it is imperative to gain insight into Black Holes. What are they doing in cosmos? How do black holes fit into T Theory? What are they? Why are they so misunderstood? **T Theory says,** 'Black holes have played the most absolute vital role in the building of our cosmos.'

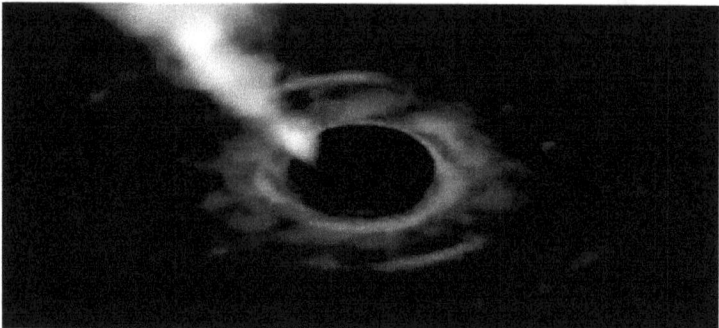

Each year, astronomers find out more secrets about black holes. At first, they were thought of as destructive eaters. Now, astronomers view supermassive black holes as control organizers at the hub of spiral galaxies. Yet, astronomers can't explain why. TT shows that the light eaters (black holes) possess 'power with purpose,' being the engines which were designed as the cosmic builders of our physical universe.'

Here is what astronomers <u>incorrectly</u> think: 'Black holes are quicksand where light is trapped forever. Or, a black hole results from a star which collapses into a dense area after a Supernova.'

TT says, 'Observations by astronomers only get a certain amount right. It's difficult for them to place the existence of black holes into Big Bang.'

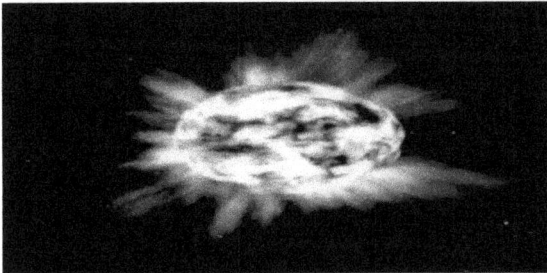

Sun goes Supernova, exploding; then it implodes.

T Theory says, 'When a star (sun) goes Supernova and explodes, the black hole which originally built that sun loses control over all the matter which it spun around itself. Hence, a Supernova occurs.'

Why then does the sun implode back into itself?

TT: 'Right after Supernova, the black hole (which was inside the dead sun) survives, now naked, empty, and free to spin fast again. The implosion is really the black hole quickly pulling escaping light back in as dense matter as the black hole begins the process of building a new sphere.'

T Theory says, 'A naked black hole always spins light into matter to build a brand new sphere.'

T Theory says, 'The Supernova sun imploding back into a black hole shows 2 things.

Firstly, the black hole was always inside the sun.

Secondly, that black hole is determined to win the battle and become the sun of the next solar system. By immediately sucking light back inwards, the large black hole gets the upper hand on all of the lesser sized black holes inside of the planets and moons in a new solar system.

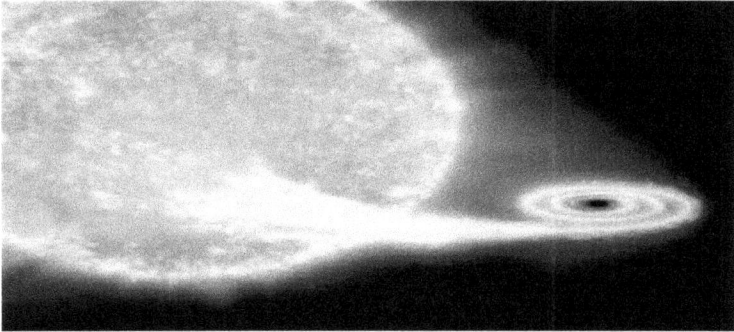

T Theory says,, The entire life cycle of a black hole is shown in detail in the Trillion series books by Ed, namely: 'Trillion Theory' and 'Trillion Years Universe Theory' and 'Black Holes Built our Cosmos'

T **Theory says,** 'Cosmos, comprised of planetoids, solar systems, and galaxies was built by machines commonly referred to today as simply black holes. This comic builder role, will be irrevocably proven by astute astronomers over the next decades. Once proven, the bigger question becomes, WHO designed black holes as cosmic builders?

T **Theory says,** 'We can take our first baby steps in uncovering the WHO and WHY of our universe by fully understanding what black holes are and how they have continually methodically completed the cosmic-building tasks over the past trillion years.

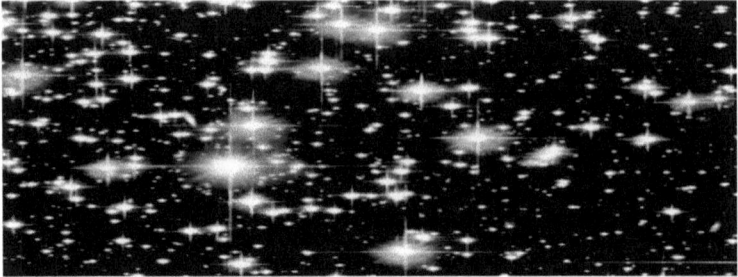

T Theory says, 'A first step in grasping T Theory is to accept the notion of a black hole at the core of every sphere. During the process of spinning fast to attract and eat light, a naked black hole builds matter around itself to then appear as a sphere. The spin of a black hole provides the spin and the projected gravity to its sphere. This is the similar case with all spheres in all solar systems. However, the rate of spin of a matter-loaded sphere, with a cloaked black hole at its core, will be slower than the spin rate of that black hole when it was empty.

This idea recognizes black holes as the builders of all the spheroids (planets, moons, suns, stars), as well as the builders and organizers of all solar systems within organized galaxies. Furthermore, the arrangement of the orbs within a solar system is the result of a gravitational battle which took place between all the many different sized black holes when the solar system was forming.'

If a person accepts the T Theory premise that a black hole is present at the core of every planet, moon, and sun in all the solar systems in all the galaxies of our cosmos; and that a 200-800 billion year old aged supermassive black hole is at the controlling central hub of every galaxy, then knowing the ins and outs of black holes becomes paramount to understanding the exciting origin, history, and operations of our cosmos.

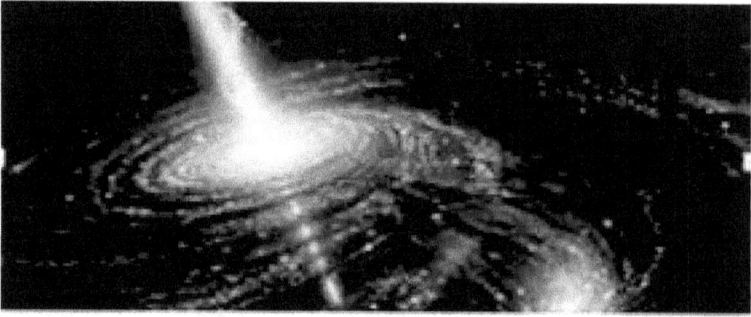

T Theory Says, 'No one has ever realized the role played by evolution in the growth of the billions of gigantic galaxies across the cosmos. Not only are massive black holes at the center of the suns which build and then recycle all the solar systems on an approximate 15 billion year calendar, but massive black holes have adapted their internal mechanisms enabling themselves to evolve into supermassive black holes at the hub of galaxies.

Read Trillion Theory books for greater detail as to how the mechanisms of black holes operate.'

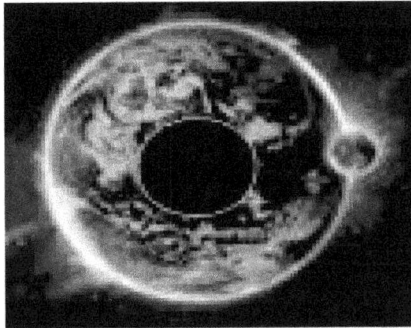

T Theory says, 'A cloaked black hole, of small to medium size range, is at the core of Earth. Prior to its building of our Earth, this black hole spun at an ultra fast rate as a naked black hole when billions of years in the past it attracted and spun light into matter, building a body to form Earth. Today, this black hole at Earth's core spins slower because of its acquired mass. Yet, it can still extend sufficient gravitational pull to hold us and other objects on the surface, and also hold our moon in orbit.'

To this day, black holes are very misunderstood. Hopefully, T Theory will usher forth a new way to view black holes, thereby building a whole new respect for their importance. Our cosmos could never have been erected without black holes as the catalytic engines. T Theory aspires to be a powerful new voice echoing for black holes.

'Weirdly mysterious.' Quite simply, what we don't yet know about black holes far outweighs what we do. Prior to T Theory, no one had been able to explain their presence. Till now, black holes were portrayed as bizarre cosmic monsters. Finally, a few astronomers agree that black holes can bring orderly organization to our cosmos. Astronomers are slowly converging to T Theory. Next, T Theory needs to get an astronomer to discover a cloaked black holes inside of a moon, planet, or star.

Here is what astronomers previously thought:
- black holes were a rare cosmic feature.
- black holes were difficult to find in space.
- black holes were a one way street for light.
- black holes could spin furiously.
- black holes all spun in the same direction.
- black holes brought only chaotic destruction.
- black holes could destructively rip apart a star.
- black holes could make a star go Supernova.

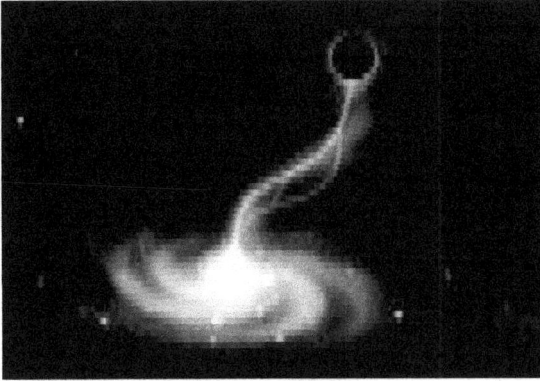

Here now are new ideas which many astronomers are discovering about black holes:
- black holes are a common cosmic feature.
- black holes are becoming easier to find.
- black holes extraordinarily exist by the billions.
- black holes are mysterious cosmic objects.
- black holes aren't just chaos and destruction.
- black holes aren't just a one way street for light.
- black holes can spin at millions of mph.
- black holes can spin in either direction.
- black holes stage the most violent battles.
- black holes also have another gentler side.
- black holes are purposeful cosmic organizers.
- black holes seem to shape our cosmos.
- supermassive black holes are at galaxy hubs.
- supermassives control and organize a galaxy.

Yet, T Theory says, 'Keep looking much deeper.'

T Theory says, 'One secret which astronomers have yet to discover is that all black holes are sphere factories, spinning light into orbs. Black holes then reside inside of the orbs for billions of years. Therein, as spinners of light to matter, black holes built the spheres existing in all solar systems. Indestructible black holes have been sphere masons and sphere recyclers for the past trillion years of cosmic history.'

To astronomers, light devoured by a black hole was seemingly lost forever. But in 1974, Stephen Hawking made his famous discovery: black holes emitted radiation. Hawking now declares; black holes exist differently than first thought; adding to his old stance that light can't escape from a black hole by admitting that light is sort of stuck or stored in a holding pattern inside a black hole.

T Theory likes that Hawking is seeing the light. TT says, 'Black holes spin light into matter. But, matter always returns to escaping light, even if it takes billions of years. A sun is an example where a tired black hole has lost the grip which it held on its matter. Each day, billions of atoms escape as light from a sun, and as the sun ages further it is destined to someday become a Supernova.'

T Theory says, 'Black holes are ubiquitous to our universe. T Theory predicts that black holes will be discovered existing all across cosmos. Already, we know of the supermassive omnipresent black hole at a galaxy hub, with up to millions of suns and solar systems under its control. T Theory further predicts that an x-large (or massive) black hole is at the core of every sun; and a smaller black hole at the core of every smaller sphere. Black holes provide the spin and gravity seen in spheres.'

TT says, 'The bigger questions become: how did black holes become cosmic builders? Where are they from? Who owns this Black Hole Society? Or, are they the active owners of our universe?

Because of the phenomenal spinning powers of black holes, cosmos perpetuates its own growth. 'Nobody has to run out to turn a crank or plug-in a charger when it is time for cosmos to recycle.'

T Theory professes the following claims:

- Cosmos, at a trillion years, is far older than the 13.7 billion years estimated by astronomers.
- Cosmos grew to its size of 73 quintillion stars, with no Big Bang in sight. (The only super big explosions are the Supernovae from aged suns).
- No to happenchance. Rather, specific scientific design is behind the origin of our cosmos.
- Black holes were designed as cosmic builders which spin light into matter to form orbs.
- A black hole (sizes vary) resides at the center of every moon, planet, star, and galaxy.
- Solar systems, and all their spheroids, comprise the galaxies of our cosmos.
- Galaxies have permanency, while resident solar systems recycle every 15 billion years or so.
- Each recycle of a solar system ups the count of stars and solar systems inside ancient galaxies.
- There's an endless energy supply (light ocean) which black holes access to grow the cosmos.
- Light, with its incredible properties, is deployed as a singular material in the recycling process.
- Black holes can self-replicate (split into 2) as a result of a Supernova and then the subsequent Obliteration of a solar system which results in all of the planets and moons melting and exploding.

More T Theory rules regarding black holes at the cores of planets, moons and suns:
- all spheres built by black holes have an upper age limit of 15 billion years. The exception is a supermassive black holes at the hub of a galaxy.
- planets/moons only survive 10-15 billion years, dependent upon the lifespan of their sun.
- all planets/moons fire-up inside their black hole core, as if destined to become a sun, but their life is cut short when their sun goes Supernova.
- they all ultimately see their matter destroyed.
- all suns eventually die, going Supernova.
- Supernova destroys an entire solar system.
- all the black holes survive the Supernova.
- black holes replicate as their sphere dies.
- black holes always get back to work to attract light to build themselves a new body of matter.
- black holes always rebuild another solar system.

Someone asks, 'How do astronomers know where to look for a naked black hole?' T Theory replies, 'All the black holes cloaked within cosmic spheres are hidden from view as they're no longer naked and empty, but rather brim-full with matter.'

TT states, 'Naked black holes, in the process of attracting and eating light, are found as follows:
- On the perimeter of space, at the ocean of light, where as worker black holes they break off chunks from the frozen ocean and devour freed light.
- Also, after a sun goes Supernova, where black holes survived Obliteration of the orbs that they had occupied in the destroyed solar system. The naked black holes all commence a battle for light to build into new spheres for a new solar system.'

People ask, 'If nothing can escape from a black hole, how can light escape our sun when T Theory says that a black hole is at our sun's core?'

Good question. How can a black hole at the sun's core ever allow light to escape. Aren't black holes light eaters?

TT answer, 'Yes, naked black holes are eaters of light, spinning light into matter. But, black holes always overeat, especially the largest black holes of a community, namely the XL black holes which evolve into suns. The weakness of an XL black holes is that its fast eating spun hot loose atoms in its core. Of all the matter which was spun by all of the black holes holes at the core of spheres in a solar system, the XL black hole at the core of a sun spun the loosest matter. This loose matter is the first to lead the charge, utilizing the Cosmic Rule, 'Light spun as matter always escapes going back to light, even if it takes billions of years.'

So, the XL black hole of a sun looses control of matter which escapes as light. This is a precursor to the time when the sun will go Supernova.'

'Always see our cosmos from a strategic point of view. Black holes always do things for a reason. They want to be biggest, with most control over lesser-sized black holes. They accomplish this by over-powering others, or by out-foxing them.

So, for an XL black hole at a sun's core, getting rid of all of its contents before any other spheres in its solar system, is a smart strategy move. The XL will get naked and be eating light before any other black holes when Supernova/Obliteration hits the solar system. XL has the upper hand to grow larger and become the central black hole in control of a new solar system. Excellent strategy.'

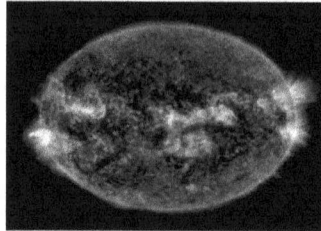

TT says, 'A fiery sun centering a solar system is the norm throughout all solar systems renting rooms in galaxies. A sun without a solar system surrounding it, is near impossible. TT will show how solar systems formed, with all of its black holes abiding by specific cosmic battle rules.'

T Theory says, 'Here are more goals of T Theory.'
- provide a much more scientific approach via a
  theory which digs deeper for cosmic truths.
- show cosmic age at a trillion years since origin.
- sell the notion that our cosmos began small,
  then grew through 15-billion-year generations.
- credit black holes, as scientifically designed
  machine-like entities, for growth in numbers   of
  solar systems and galaxies within cosmos.
- to see TT displace Big Bang as a paradigm.

Proclamation from T (Trillion) Theory (TT).

Small changes give rise to large consequences
and incremental changes turn our understanding
on its head. We need to undertake a far greater
cosmic introspection by performing forensics on
black holes, determined to find all of the truths.

Now that you have been getting a glance at Trillion Theory (TT), you may have many concerns and questions. So for a moment, let's step back: T Theory is the brainchild of Ed Lukowich.
He was disenchanted with a faulty Big Bang. Ed began formulating T Theory back in 1998; Over the past years, revisions took place for the best way to present this new theory. Then, inaugural publications were in 2014, 2015, 2016.

When your read the cosmology books of Ed Lukowich, you will immediately note the focus on the role that black holes have played in the construction, recycling, growth, and control of the cosmic solar systems and galaxies. T Theory demonstrates the outer features of black holes which allow them to be the dominant gravity players throughout the cosmos. T Theory also goes inside black holes, attempting to unveil the internal structures which give black holes the power to build matter by spinning light.

Later, T Theory also uncovers the one great strategic weakness built into black holes.

Albert Einstein laid the theoretical foundation for the existence of black holes by predicting that light would bend when nearing a sphere's gravity. But, we are only beginning to know black holes. Daily, astronomers take new looks at black holes. T Theory hopes they find evidence to support TT.

Till T Theory, no one had surmised the existence of a black hole living for billions of years at the core of every sphere. Till TT, no one had even an inkling to relate black holes to the building and to the recycling of those cosmic spheroids.

T Theory says, 'Black holes are *sphere factories* hard at work to build and recycle cosmic spheres. Then, via their gravity they provide organization and control as seen in solar systems and galaxies.'

Till Trillion Theory, black holes were never seen as having a definite duty. No one, till TT, had ever rationalized that black holes do have a true higher purpose as the ultimate cosmic builders.

Till Trillion Theory, the idea of a black hole at the center of every spheroid was never considered.

Till TT, no one ever surmised that light escapes a black hole after entrapment of billions of years.

Till TT, no one theorized how the eventual escape by light from a black hole (or a sun) is part of the recycling process of the sphere of our cosmos.

Till T Theory, no one theorized that black holes are actually alive with strong animalistic appetites to devour light and hold it tight as matter.

Till T Theory, no one theorized that the goal of a black hole is to become the largest and strongest orb capable of dominating smaller spheres; and able to be the dominant builders of our cosmos.

Till T Theory, no one theorized that there was an entity living at the core of every cosmic sphere. An entity which is a survivor. A naked black hole survives a Supernova of its sun, while other lesser sized black holes survive the Obliteration of their solar system. Black hole survival, plus an ability to replicate, grew our cosmos to 73 quintillion stars.

In today's world, astronomers and scientists are recalibrating their respect for black holes, seeing all black holes as constructive builders, not just destructive monsters. Penn State astrophysicist Yuexing Li had this to say about black holes, 'Now we see that black holes were essential in creating the universe's modern structure.'

If readers can accept the radical concept that a black hole spins light to matter, and then accept the equally radical concept that a black hole resides at the core of the sphere it built, then the next step is the most radical for acceptance. This step answers the question, 'Where did light come from in the first place as a feeder for a black hole?

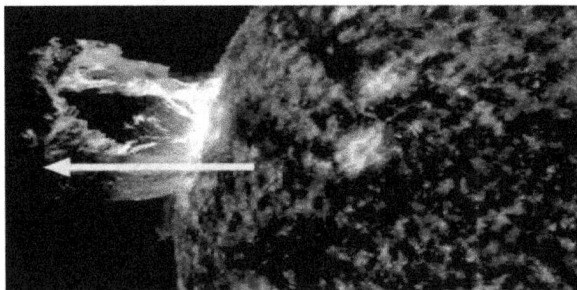

T Theory asks an upfront chicken and egg type of question, 'Which was first, light or a sun?'

Everyone, following their eye would say, 'That's easy, we see our sun and then we see light coming from our sun, obviously a sun was first.'

But, TT disagrees. T Theory shows that before there were suns, stars or space in our cosmos, there was only light. TT's, Light Was First theory.

A **reader replies,** 'Then, if light was first, this light was traveling through space and then eaten by a black hole to form a sphere of matter?'

T Theory responds, 'Today, a black hole can eat light from a sun (star). This light travels through space after it leaves the sun.

But, T Theory says, 'Space didn't yet exist at the origin of our cosmos. Remember, TT says that light was first.'

T Theory claims, 'Before there was space throughout our cosmos, first there was only LIGHT. So, light was not born into an area called space, for light existed well before space.'
This is quite the statement by Trillion Theory, which demonstrates how light preceded space.

Our universe is tricky; often it fools our eyes and our reasoning. TT says, 'We as humans tend to think that since the contents of our cosmos sit in space, that space was there first. Example: In Big Bang theory, supposedly matter exploded into space which was already there as an empty area, ready and waiting to accept matter. We see space as the box which our universe was placed into.'

But, T Theory says, 'NO! Space was never sitting there waiting with open arms, like a box to be filled, waiting for cosmos to be created. For, T Theory states that space was not ahead of light, but rather, space is that huge vacated area left behind where black holes spun vast amounts of light into ultra-dense compressed spheroids.'

Thus, TT says, 'Picture an earliest cosmos as tons of light (along with its properties such as weight) spun by black holes into ultra-tight packs called spheres. Once this task was done, all of the light was inside of the spheres, while space is the vast, empty, void, weightless, cavern left behind.'

TT says, 'At the cosmic origin, all that existed was a never ending ocean of static frozen light.'

TT says, 'Think of the ocean of frozen light as the first strategy step by a very savvy strategist.'

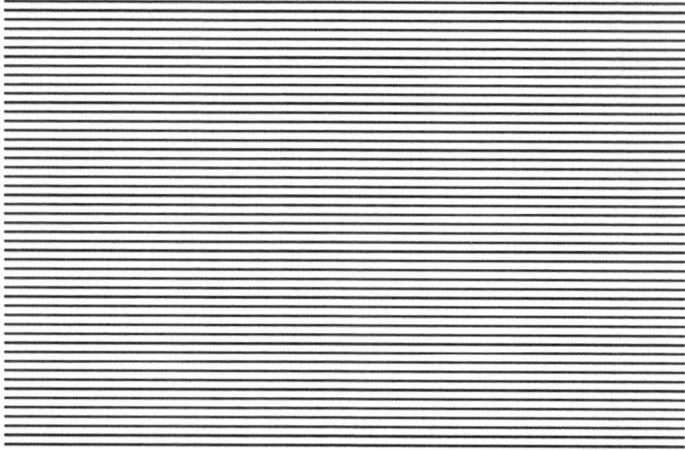

T Theory says, 'LIGHT FIRST was all that existed at the beginning of our cosmos a trillion years ago. An endless ocean of FROZEN light. TT claims that cosmos was born from this playing field; an endless ocean of frozen static non-moving light.'

T Theory claims, 'Frozen light is totally foreign to us, as no one had ever seen it. Then, in 2014, physicists in their lab froze a light ray for the first time ever, slowing its speed to zero. This frozen static light was a great find to assist TT.'

A trillion years ago, at the origin of our cosmos, the frozen static strands of light in the light ocean occupied the entirety of all that existed in this realm. So at the outset, before our cosmos even began, all the material necessary to build our cosmos, namely light, was set into place, ready to be deployed. These massive amounts of frozen light strands in the light ocean weighed plenty as they also carried an array of traits, namely weight, heat and the light spectrum in light's toolbox.

The light strands waited in their frozen state in the light ocean, awaiting action – precipitated by the introduction of a spinning naked black hole.

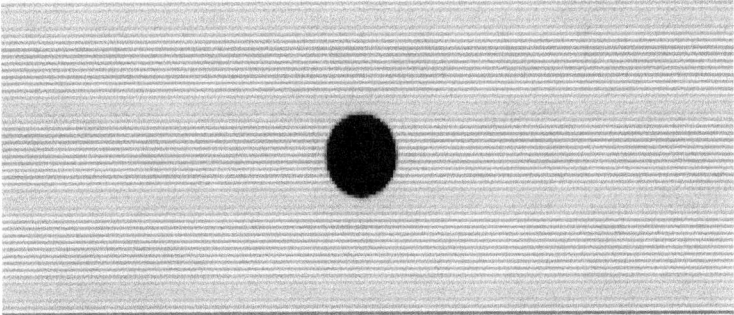

T Theory says, 'The opening cosmic event began a trillion years ago, with the introduction of the first naked black hole into the endless ocean of static non-moving frozen light strands.'

This inaugural black hole spun at great speed as it attacked the ocean of static frozen light. With its churning spin, the black hole broke chunks of the frozen light away from the ocean. The chunks then separated into rays of moving light. A small empty space cavern replaced the depleted area. Instantly, the freed light accelerated, ricocheting off the walls inside the cavern. Around this space, the walls of the light ocean held firm, still frozen in place. As more light was freed from the wall, it too quickly sped to move at light speed within the expanding space cavern. Most of the rays were dragged by the gravity pull of the black hole towards it, to be spun into the first atoms of matter. Continuing its attack on the walls of the ocean, the black hole broke off more chunks, freeing more light; spinning that light into matter.

The bodily size of the first-ever naked black hole grew dramatically as it ate more from the frozen light ocean. Each acquired light strand spun to fill the body destined to be the first cosmic sphere.

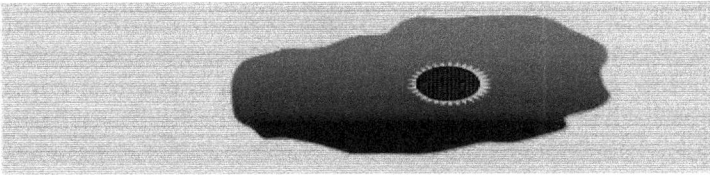

The diagram shows the scene at the cosmic origin after the first-ever black hole ate light to form a sphere from the ocean of light. The outer perimeter is a frozen ocean of light with hard interior walls, like the inside of a mine. The blackness inside the ocean is a cavern of space.

The sphere (in the diagram's center) was so densely compacted that it left behind a large cavern of empty space inside of the light ocean. Area ratio of orb to space was 1 to a million.

The space cavern was cucumber shaped since the black hole had eaten from the frozen ocean along an equatorial plane extending outwards along its gravity field. Note: Space grows vaster as more black holes spin light to matter.

Originally, the naked black hole spun fast, but as it filled, its spin rate slowed, encumbered by the tons of heavy atoms of matter it had spun.

The cosmic growth rate was dependent on how quickly more black holes entered the light ocean.

This first sphere spun on the axis of the black hole at its core. But, its life had been transformed. The full black hole went from spinning light to matter, to a long holding pattern as it attempted to maintain control of its tons of matter. Spun light was jailed for billions of years, till the oldest atoms of matter loosened to fire-up deep in the black hole's belly. This fire turned to lava which expanded, creating fissures to squeeze through and then volcanically reach the orb's surface.

But suddenly, a 2nd black hole was introduced into the realm of the young cosmos.

The 2nd black hole broke more chunks from the walls of the light ocean as it used its fast spin rate to gnaw away. It captured many rays, spinning them into matter inside its core; while, some other freed rays moved at light speed through an an expanding space cavern. Also, this 2nd black hole caused interaction with the first black hole.

The first and 2nd cosmic black holes spent eons competing with each other, an interaction which set the initial rules of engagement inside cosmos.

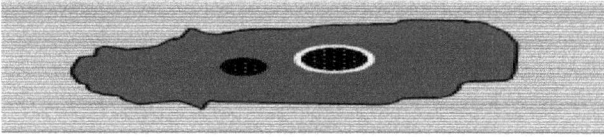

The mass of the first black hole used it greater gravity to pull the new black hole towards itself. The first tiny cosmic system was formed with one larger sphere at the hub and one smaller orb in orbit. The new orb, from its orbit, consumed light from the larger orb which was fast losing control of its matter which departed its surface as light. Thus, the first cosmic sun resulted. A sun which ironically held its eater in orbit, while the eater went through the process of destroying it.

As the newer black hole ate from the new sun, it too filled up its belly and began to form a body. But, it never knew what was about to hit it. The sun lost control and Supernova exploded. Much of the matter spun by both black holes melted away, and scampered through space as light. All that remained of the first-ever solar system was the two surviving naked black holes.

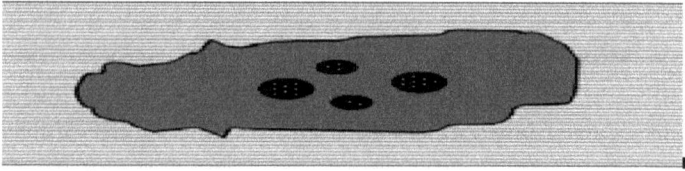

The power of Supernova and of Obliteration split both of of the black holes at their axis, each of them thereby replicated from one naked black hole into two; 4 naked black holes resulted.

Interaction began between 4 black holes. One of the two larger would center a new solar system.

With the 4 naked black holes, interesting events occurred. They all competed to attract light with their gravity and spin that light into matter to fill their gut and build a body. The two largest black holes ate the quickest, spinning their light into atoms of looser matter. The best eater grew the fastest and thereby took the hub at the center of the new-forming solar system. The two smallest black holes spun denser packs; they grew slower, finding orbits around the larger orbs.

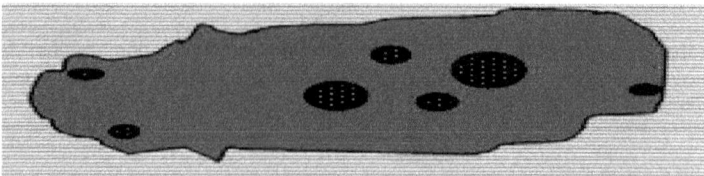

On the outer limits, at the end of space, at the wall of the ocean of light, brand new naked black holes were being introduced into the cosmos. These worker-bee black holes broke chunks free from the frozen ocean walls, further increasing the size of space, while freeing light to travel through space and be utilized by the 4 black holes which were building a new solar system.

TT definition, 'Space is a gigantic cold empty weightless black cavern within a frozen ocean of light. This cavern resulted from the action of naked black holes cutting into the walls of the light ocean, breaking off chunks of light and spinning that freed light into cosmic spheres.'

Equate cosmos to mining: Miners burrow into walls of a mine, reaping ore, while extending the overall empty area inside of the mine. In cosmos, black holes eat into a light ocean, taking light to spin into spheres; leaving a huge space cavern.

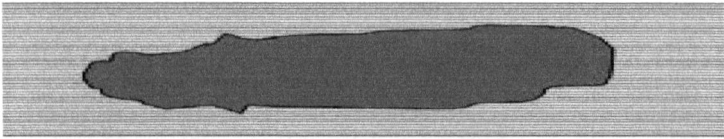

TT says, 'Today, if we traveled to the very outer edge perimeter of cosmos, we'd cross trillions upon trillions of miles of space. Eventually we'd come to the end of the space cavern. But, also we'd run smack dab into the hard surfaced walls of the frozen ocean of light surrounding the outer perimeter of space. This ocean holds a ready endless supply of new light to be resourced by naked black holes and spun into even more cosmic spheres. At the ocean walls, we'd find new worker-bee naked black holes, busy breaking off chunks of light, freeing that light to travel and supply a cosmos which is forever growing in its number of solar systems and galaxies. These worker-bee black holes would capture some of the freed light and spin it into matter to form themselves into a sphere. These actions further expand space.'

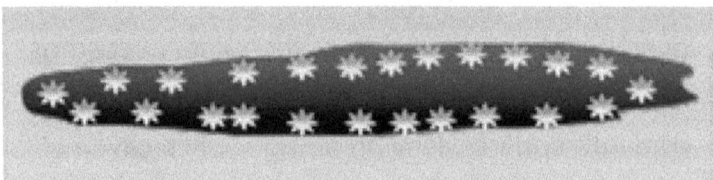

T Theory says, 'Black holes, as cosmic builders, determined the size of space. The reason that the distance between the stars is so staggering is because of the megatons of matter packed into spheres when compared to the billions of miles of void space left behind. This huge cavern of space is so vast while spheres occupy so little. Space takes 99.9% of the cosmic area; while galaxies, solar systems, stars, planets, moons, gaseous clouds, comets and asteroids, take up less than 1%. That is why distances between galaxies is so vast. Comparatively, matter is so densely packed into orbs, that one atom of matter spun from light leaves one trillionth that amount of voided space behind.

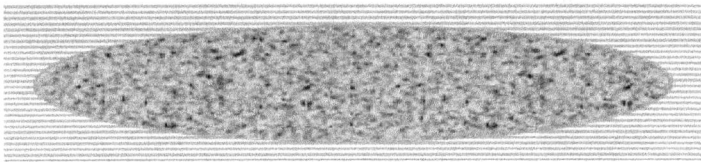

Today, inside of the endless light ocean, cosmos has grown to be trillions of light years across a cavern of space. Tightly compressed is all of the matter taken from the light ocean and spun by black holes into the spheroids inside of galaxies.

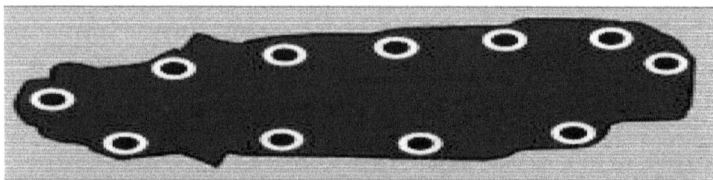

To summarize, T Theory shows that the outer perimeter known as the frozen ocean of light provides cosmos with an ongoing supply of fresh energy to grow a cosmos to infinity. As well, the space cavern continually grows larger along a linear plane each time a new naked black hole eats into the wall of the ocean of light. Once freed, some of this freed light travels through space to supply the black holes, solar systems and galaxies already inside of cosmos. The size and number of total spheres, and total solar systems continually increases, as does the total size and the numbers of cosmic galaxies.

As you read about T Theory, it becomes evident the emphasis which TT places onto light and also black holes. Together, they share the spotlight. Here is a closer look at wondrous light as never examined before. **TT says,** 'If physicists can spin light into atoms of elemental matter at their lab, that astonishing feat irrevocably proves TT.'

T Theory says, 'The way in which we view light is about to undergo a huge change. No longer is light just able to come to us from our sun to shine and reflect. We know that light bends in a rainbow when passing though water, refracts through a prism, jumps from a fire, is harnessed as electricity, flashes in lightning, and escapes during a nuclear explosion; but T Theory now affirms, light can go beyond a bending point to be spun into atoms by a black hole. Once spun, light spins for as an atom for eons looking for its opportunity to escape and travel as light.'

How can light do all of these magical feats?

T Theory says, 'We have yet to uncover all the incredible properties and talents of light.'

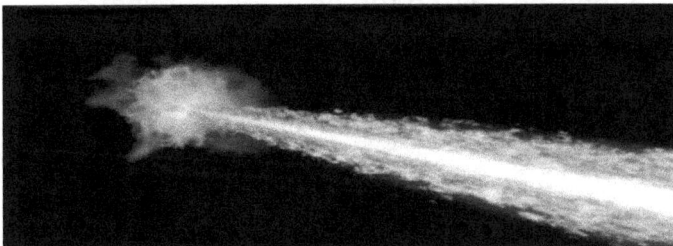

Trillion Theory places LIGHT as the material at the forefront of our cosmos. Humble light likes to hide its uncanny talents, which we must seek to find. But, with each new discovery, we learn more to add to light's phenomenal properties.

T Theory says, 'Light, the single material deployed to construct cosmos, exists in 3 basic ways:'

- FIRST, as frozen static light as existed at cosmic outset a trillion year ago, and still exists today on the outer perimeter surrounding cosmos.

- 2nd, as atoms of matter where light is captured by a black hole, imprisoned for billions of years inside of a spinning atom of matter.

- 3rd, as straight line light traveling at light's top speed . This free traveling light is either broken away from the frozen light ocean on the cosmic perimeter, or escaped from its imprisonment as matter from a star.   (Note: As well, we know of ways to store light in forms such as electricity to then be utilized for heat and illumination).

T Theory says, 'Light is an indestructible material. It can be spun into atoms of matter, then spin inside that atom for billions of years till escaping back to light to travel through space. Light is vital in formula $E=MC^2$. Today, there are 73 quintillion stars, because light is a superb recycling material.'

TT says, 'We take light for granted, but don't. Because, light's properties are very multifaceted, as seen with light's beautiful color spectrum.'

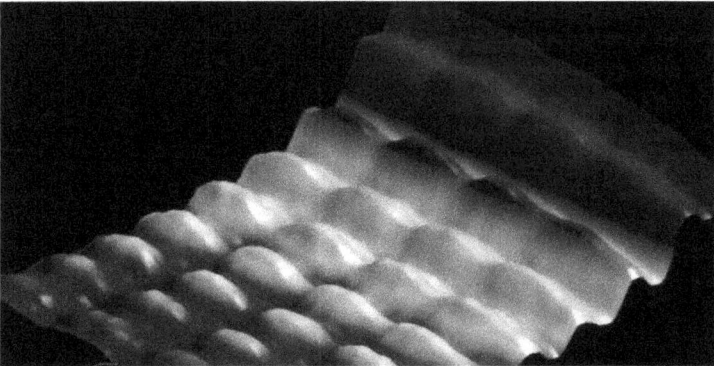

T Theory says, 'Light First. Before there was space, orbs, solar systems, or galaxies, there existed the endless frozen ocean of light.' Scientific genius built the frozen ocean of light. Each strand of light ready to be freed from the ocean, ready to move to the speed of light, yet available to be spun into atoms of matter.'

It's paramount for researchers to uncover all the secrets of light as the fundamental single material that was deployed to construct our cosmos.

Following is 4 light features: 1 and 2 are already known by physicists, while T Theory adds 3 and 4.

Feature 1: Speed.

♦ Light is the fastest cosmic traveler that we know of. It speeds at its constant, referred to as 'c' is186,282 miles per second, in a vacuum or in space. (5.9 trillion miles in a single light year; or, one parsec in 3.26 light years).

Feature 2: Change Speed.

♦ Light's speed stays at a high constant unless slowed by a gravity force. Light is slowed and bent when encountering a black hole's strong gravity. At a Gravitational Station, if the black hole is naked it spins light into matter. If the black hole is cloaked inside an orb, it can only attract and bend, but not spin light to matter.

Feature 3: A Carrier. (TT Says).

◆ Light carries its own tool box. Inside of the tool box are the properties light will need if ever gets spun into matter, namely: heat, weight, electromagnetic spectrum, plus the entire array of extensive subatomic properties.

Feature 4: The Ability to Spin into Matter. (TT).

▪ Light slows when grabbed by a black hole.

▪ Light has flex, it can bend, coil and spin.

▪ Light spins forming the body of an atom.

▪ Light can form a tight or a loose atom.

▪ Light spun continues to spin for eons.

▪ Light's weight is shown in an atom's weight.

▪ Light spun into more atoms will also form a body (sphere) around a cloaked black hole.

▪ Light is held billions of years by a black hole.

▪ Light trapped inside an atom wants to escape.

▪ Light is always successful in its escape.

▪ Light, when it finally escapes, is free to travel.

TT says, 'An immense amount of light is needed to spin the atoms of the body of matter around a sphere. Formula $E=mc^2$ calculates the huge ratio amount of light energy which goes into the relatively small amount of formed matter.'

T Theory says, 'As light is pulled into a black hole such as in the image above, that is the beginning of the 15 billion year or so cycle for light to go from being able to travel as straight line light, to billions of years of imprisonment as an atom of spun matter. Then, finally escaping one day from a spheroid as a light ray making its departure.'

When it comes to the type of an atom which light is spun into by a black hole, T Theory admits that is difficult to know at present. But, T Theory does offer an analogy as follows.

Think of light as yarn. Knitting makes wonderful garments from yarn. The knitter can use different colors of yarn, various types of knots, a variety of patterns, and different loops, to create a woolly scarf, toque, sweater, mitts, socks, or a blanket. They all look different, but underneath they are all made from one material, just yarn.

In reverse, the knitter can take each garment and unravel it back to original yarn. Regardless of disguise, in or out, garments are just yarn.

Light has unique properties which allow it to be spun into a whole number of different atoms. On Earth, there are well over 100 pure elements, and every one of these elements has a specific atomic arrangement. There may even be other elements, even on other planets, waiting to be discovered.

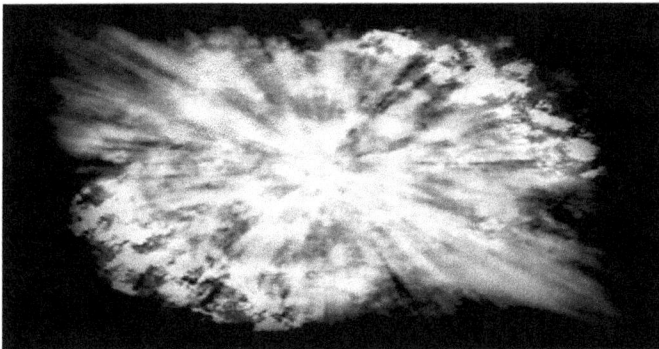

When atoms of matter unravel, many tons of light escape (along with the properties of light). Taking that thought even further, when the atoms of a star unravel (during a Supernova), the amount of energy released is incredible (Formula $MC^2=E$). The amount of energy volume which escapes back to straight line light would be staggering to a zillion zeroes. On the opposite side, the amount of light originally spun into atoms of matter to make a planet or moon or sun is uncountable.

Most of what TT proposes is new to cosmology. Certainly an idea of light as the singular material used to build our cosmos is totally foreign. Also neo TT shows space as this cavern resulting from black holes spinning light into matter from the light ocean. Just as radical is the idea of a black hole being present at the core of each sphere.

So you ask TT, 'Why are there black holes in our cosmos? Why does TT say there are billions?'

T Theory replies, 'Light, as the cosmic material, does require an action machine to be the catalyst which exposes all of light's scientific properties. It is black holes that are the cosmic action engines.'

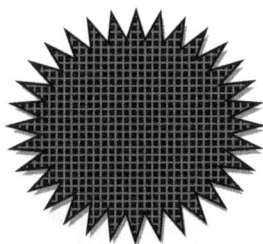

T Theory says, 'Sometimes an astronomer is lucky enough to spot a naked black hole before it has a chance to attract light. Too late to the party, and the astronomer gets glared-out by the brightness of light entering the black hole.'

Black Hole Structure

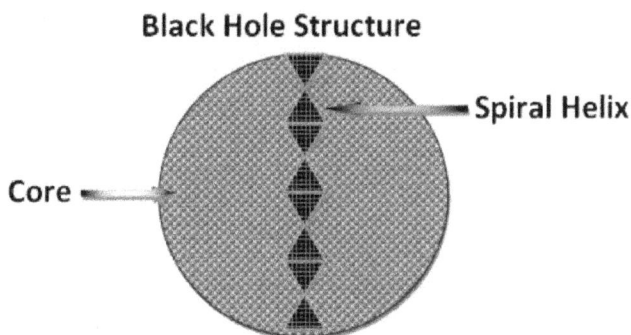

Spiral Helix

Core

T Theory says, 'Unless we are able to get super close to a naked black hole, or capture one, we might have to be satisfied with only guessing at its interior. T Theory makes an attempt here.'

T Theory says, 'Our cosmos often drops us a clue so we can discover the next great thing: for example, a rainbow suggests that light bends and refracts, but that's only a clue to make us find out how light spins to form matter. If we look for clues as to how a black hole spins and then extends a gravity, T Theory suggests that we examine the inner bowels of a spinning top.'

T Theory says, 'For clues to how a black hole's body operates, TT studies a pumping toy top. A person repeatedly pushes the handle of the toy down to impart spin as it pivots on it axis on the floor. Inside the toy, the handle connects directly to a twisting helix rod which transfers its rotation to the body, thereby making the body spin.'

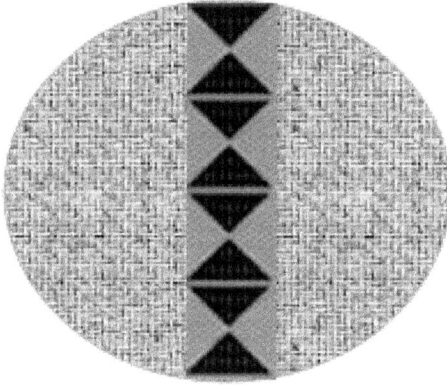

T Theory suggests that if we entered the inner workings of a naked black hole, we would find billions of tiny compartments, all ready to house atoms of matter to be spun from light. Also, we'd experience a turbo-charged spin, the fastest in cosmos. At the center of the black hole, running north-south, we would find a spinning helix rod which can simultaneously deploys elasticity to lengthen and then shorten itself as a means to perpetuate its continual spin. The turning of this central spindle transfers torque to spin the entire body. The spin of the body is so dramatic that a so-called draft (gravity) extends out past the equator of the black hole. In essence, this naked black hole, spinning at an unabated super speed, is ready to capture and spin light into matter.

Inside a naked black hole, processes occur. Light, pulled inwards from the event horizon, is taken deep into the core. The first attracted light goes all the way into the spiral helix. In TT, the laws pertaining to black holes show that the spiral helix will fill first, able to grow its length and girth. This is evolution. Black holes are programmed to become as large and powerful as possible. They are in competition against all other black holes.

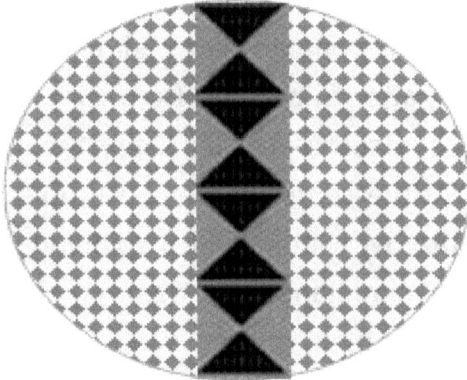

Next, the black hole will spin light into matter to fill all the compartments of its main interior body. With this fill up, every compartment utilizes its elastic feature to stretch and to subdivide. With filling, stretching, and subdividing techniques, the black hole possesses components to grow larger.

'If T Theory is right, within every black hole there exists the mechanisms to spin light into matter.'

In this author's novel 'The Trillionist', a man-made machine called the Quantronix emulates a black hole by spinning light into elements. The Quantronix's long spout protrudes out the top of a domed building, pulling light down to its belly to be spun into atoms of precious new elements.

'The spinning of light into matter by a Quantronix machine would prove black holes build matter.'

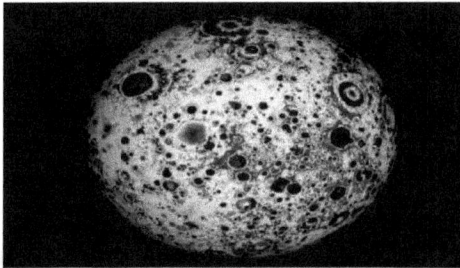

TT says, 'An integral part of uncovering cosmic mysteries, will be for astronomers to determine how a black hole accomplishes the building of an orb around itself. Such tasks has been done by black holes for a trillion years. This still occurs all across cosmos as new solar systems replace old solar systems whose star went Supernova and left behind a graveyard of surviving black holes.'

T Theory says, 'A naked black hole uses its turbo spin rate to project a field of gravity around itself which is used to attract light, spin light to atoms, and thereafter hold that matter around itself.'

The shape of the gravity projected by a black hole is as an equatorial gravity field which exerts its force along a disc-like plane past the equator. This flat gravity field exists around: all naked black holes; the supermassive black hole at the hub of a spiral galaxy; all suns which center solar systems; all planets/moons built by black holes.

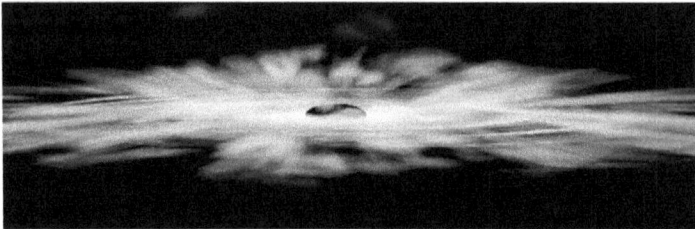

TT says, 'When black holes were designed, built-in GRAVITY was the super-glue providing the black hole with its power. This gravity is also the sheriff which enforces the cosmic Rules of Engagement, between all black holes.'

TT examines how gravity really works; where it comes from; and how its control features keep our cosmos in tact. (Astrophysicists don't as yet follow TT and place a black hole at the core of every sphere. Thus, their explanation of gravity will be less accurate than provided by T Theory).

Here is a physicist's gravity definition: 'Gravity is a force attracting things towards the center of the Earth, or toward any other physical body such as a planet, moon, or sun having mass.'

Physicists credit the mass of a sphere for its gravity. Unfortunately, they fail to follow the lead of T Theory in recognizing that a black hole residing at the center of a spheroid is the reason the spheroid spins and extends gravity.

TT says, 'The gravity of a sphere, such as a planet or moon or sun, is produced by the powerful spin of the black hole which resides within the sphere.'

Examine where gravity is found in cosmos.

- gravity on Earth keeps our planets contents on the surface from flying into space.
- gravity on Earth keeps our moon in orbit.
- gravity is seen in all planets and moons.
- gravity sees our sun as the largest player in our solar system, holding all smaller orbs in orbit.
- gravity is seen when one large solar system can hold a lesser sized solar system in orbit.
- gravity is greatest in gigantic spiral galaxies where millions of solar systems are held in orbits around a supermassive black hole.

TT says, 'The great force throughout our cosmos is gravity, which comes from black holes.'

When naked, a black hole turns its fast spin into gravity to attract and spin light into matter. After getting full, a cloaked black hole within a sphere spins slower using its gravity to keep contents on the surface and to hold other spheres in orbit.

In all solar systems, the sun at the center of the system uses its gravity to hold all the other orbs in orbit. The drag from these spheres causes the black hole at the core of the sun to tire quickest, being the orb losing control of its contents.

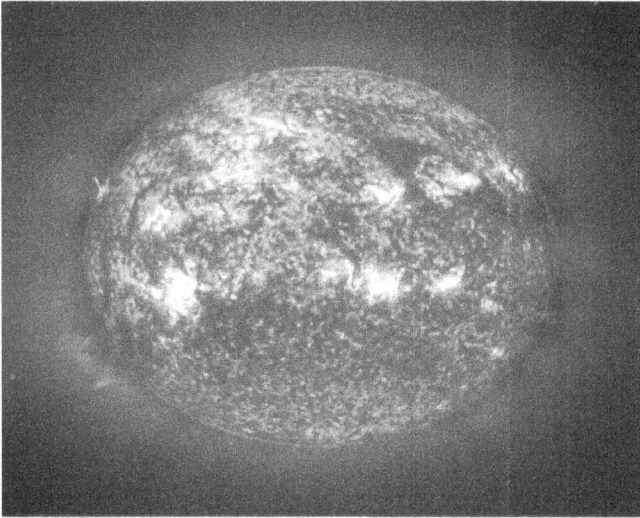

T **Theory says,** 'Of all the spheres within a solar system, the sun which is the largest, possessing the strongest gravity, should as the dominant player, be the last to ever lose control of its contents. But, there are cosmic gravity rules that work against our sun. Back when our sun's XL black hole was naked, it ate the most light and spun that light into the loosest of atoms. Now that our sun has to focus its gravity on holding 8 planets and 168 moons in orbit, it is becoming tired. The gravity pull from all of the other orbs tugs away at our sun. This has hastened our sun to lose control of its matter. Each day we see more atoms escape the sun's gravity.'

T Theory says, 'To further understand GRAVITY as the super-glue holding the various parts of our cosmos together, T Theory takes a closer look at how black holes create this gravity.'

T Theory Says, 'When a black hole sucks in light, that light enters the black hole via a side door. When a black hole spins around its axis, the greatest gravitational pull is on a plane extended sideways from its equator. In the diagram below, T Theory depicts tentacle-like filaments around the black holes equator helping to pull the light inwards through the

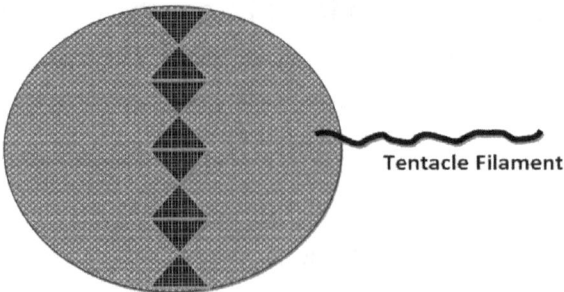

Tentacle Filament

Anecdotally, it was Stephen Hawking who recently described black holes as having soft hairs to aide in their process of drawing light inwards. T Theory takes this a step further stating that black holes have trillions of these hair like tentacles surrounding their equator.

This equatorial flat plane is totally discernible with the formation of the body and arms surrounding a supermassive black hole at the hub of a galaxy. This is the same type of flat disc-like formation we see with the rings around Saturn. T Theory shows that this flat plane of gravity around a centralized black hole is prevalent throughout our cosmos.

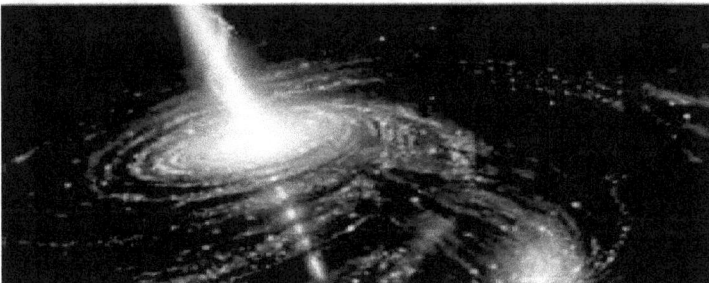

TT **redefines gravity.** Astronomers say it is caused by the spin and the mass of the spinning object. But where did spin and mass come from? TT **says,** 'All initial spin behind gravity is always provided by spin from a black hole. Furthermore, all mass comprising spheres came from naked black holes spinning zillions of tons of light into matter.'

T Theory takes us around the outside of a black hole. An empty naked black hole spins at a tremendous rate, creating a gravity field around itself. In the Event Horizon, directly surrounding the equator of the black hole, all light is drawn inwards and spun into matter. Further outwards, the naked black hole still has plenty of gravity to pull more light into its Ergosphere where it can also hold other black holes or spheres in orbits.

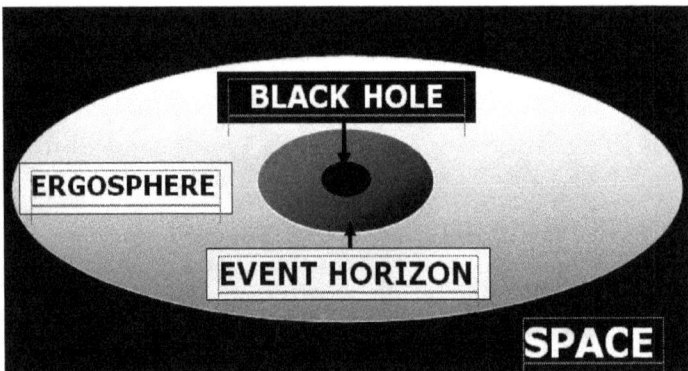

Now, what happens to its gravity when a naked black hole overeats becoming a solid planet or moon. T Theory says the gravity is still there, but just in a less aggressive nature. The body, which the full black hole now centers, spins at a much slower rate. The purpose of gravity changes from spinning light into matter, to simply holding that matter from immediate escape. Still, the slower gravity has the power to hold orbs in orbits.

Gravity is something which we on Planet Earth simply take entirely for granted. We also take for granted the spin of Planet Earth on its axis. If we someday prove a black hole resides inside of Earth to provide gravity and spin, that moment would irrevocably prove T (Trillion) Theory.

In the depiction, T Theory has placed an average sized black hole at the core of Saturn. T Theory says, 'The rings surrounding Saturn show that the black hole centering the planet rotates on its axis while extending its gravity outwards along a flat equatorial plane. The rings appear like a frisbee disc surrounding Saturn's equatorial plane.'

T Theory says, 'Gravity is 3 things, depending upon the situation which faces a black hole.
- Gravity is the strong gravitational pull which an ultra-fast spinning NAKED black hole extends outside of its core when it is consuming light.
- Gravity is the strong gravitational pull which a CLOAKED black hole extends to and out past the surface of the sphere which it built around itself when it spun light into matter.
– Gravity is the strong gravitational pull which a visible SUPERMASSIVE black hole extends out to all of the suns and the solar systems inhabiting the galaxy which it controls from its hub.

T Theory states, 'Gravity existing throughout our cosmos is supplied by and OWNED by all the black holes in control of our cosmos.'

The image shows a resident black hole (size small) spinning inside of Earth. The black hole's gravity carries along all the matter of our planet. Gravity from Earth's black hole extends far out past the surface to also keeps our lunar moon in orbit.

A following image depicts the resident black hole (XL size) spinning inside of the core of our sun. The XL black hole is losing control of its contents, but its power gravity and large mass still hold the entirety of our solar system in solar orbits. (Note: Larger suns have a massive size black hole).

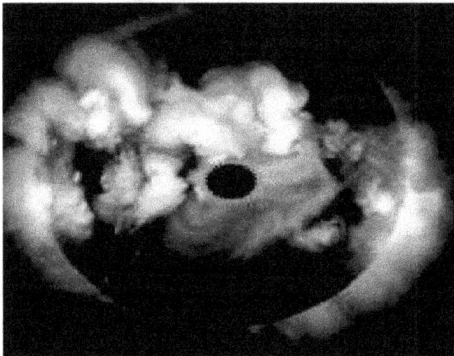

TT has talked about light, black holes, spin, and gravity, plus how all these play a major role in the billions of solar systems and galaxies which seem to fit so perfectly together.

Light is perfectly designed to spin into matter while carrying its magical properties along.

Black holes seem the perfect power engines to spin light into atoms of matter.

Black holes are perfect quiet residents to have at the core of a planet, moon, or sun, or galaxy.

Black holes supply the perfect axial spin which is seen in all of the spheroids of our cosmos.

Black holes, via their spin, provide the perfect gravity as seen around all of the spheroids.

Gravity becomes a perfect glue to hold a solar system together as well as a galaxy together.

Gravity sets the cosmic Rules of Engagement perfectly between adjacent naked black holes as well as the rules between adjacent spheres.

Light is designed with a perfect patience, as it can be locked-up as matter for billions of years, waiting to someday escape to travel space.

Light and black holes are designed as perfect indestructible recyclers within our cosmos.

TT says, 'Astonishing attention to detail is seen in the design and construction of cosmos.'

Next, T Theory will focus on the phases occurring within one 15 billion year cycle in cosmos. Light, black holes, matter, spin and gravity will all be front and center. TT contends that there have been about 67 of these approximate 15 billion year recycles by the solar systems of our cosmos. Not all solar systems recycle at the same instant, as some suns might be going Supernova at the driver's seat of their solar system this minute, while others may take extra billions of years to die in their particular domain. Regardless, they all always get rebuilt ever-larger by black holes.

T Theory says, 'A black hole follows a sequence of major phases as it builds and recycles an orb over one 13-15 billion year era. During an era, a naked black hole will build a sphere around itself, then it will live cloaked inside of that sphere, while deployment of its gravity holds the orb together. The end of the era commences when the black hole loses control over the matter of the orb; this matter escapes into space as light. After all its matter escapes, the black hole at the sphere's core is re-exposed to being a naked black hole again, completing its entire cycle.'

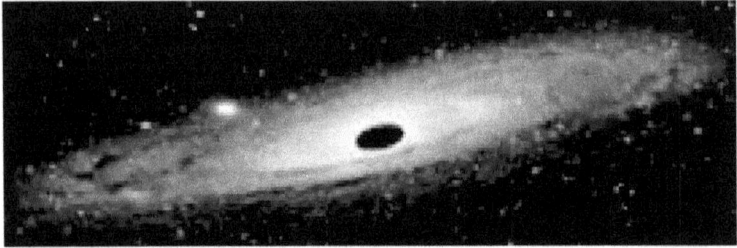

Theoretically, every black hole which spins an orb of matter around itself should have a shot at a full 15 billion year life cycle. However, few spheroids survive the full cycle because the fiery sun of their solar system goes Supernova too early.

It is worth mentioning here that a supermassive black hole at the hub of a galaxy has evolved to the point where it doesn't live by the age-limiting 15 billion year rule placed onto lesser sized black holes. Supermassive has been a game changer. It superseded the rules, evolving, thus allowing it to grow more gigantic by adapting its internal spiral to jettison the light which it eats. Thereby, a supermassive lives hundreds of billions of years to control its galaxy. During this time, all of the solar systems, which rent rooms inside the galaxy, must live by the 15 billion year limit and recycle. Throughout our cosmos, suns going Supernova at the center of solar systems are the weak link; they force their solar system to die and recycle.

Here are 5 major phases in a 13-15 billion year cycle (one era generation) for a black hole and the sphere it builds within a solar system.

Cycle starting point is the building of a sphere:

1 EATING LIGHT BY BLACK HOLE (short phase).

2 LOCKUP INSIDE A SPHERE (longest phase).

Cycle mid-point is the loosening of a sphere:

3 LOOSENING OF MATTER (next lengthy phase).

Cycle end-point is death and replication.

4 SUPERNOVA AND ESCAPE (short phase).

5 REPLICATION REPRODUCTION (short phase).

For smallish black holes, their capabilities only allow them to build a small solar system sphere.

For the largest black hole of a group, it can grow into the sun centering a solar system.

For a supermassive black hole at a galaxy hub, it has evolved superseding the 15 billion year rule; a supermassive's age can be 800 billion years.

Unfortunately, our short human lives relative to cosmic time provide us limited access. Mostly, we only see a Lockup Phase on Earth and planets and moons, and Escape as light leaves our sun.

In showing the major phases in one 15 billion year cycle, we can apply similar phases going back a 1,000,000,000,000 (trillion) years to a time when the very earliest naked black holes ate from the ocean of light and formed the spheres in the earliest solar systems.

We can also flash back just 5-6 billion years to the time when naked black holes fought for light to rebuild our solar system on the spot where a Supernova had Obliterated a previous 13 billion year old solar system.

As well, we can study the general make-up of our solar system to learn the applied rules for building a solar system: we are 8 planets, 168 moons, but only one hot fiery sphere (our sun).

T Theory phases take us through how naked black holes build spheres and then how these spheres end up in organized solar systems and galaxies. Then, how these spheres do ultimately die for the rebuilding of newer solar systems.

Phase 1: Eating of Light by a Black Hole.
Short phase. Elapsed time, hundreds to possibly thousands of years. Blink and you might miss it.

The first assignment of any naked black hole is to attract, capture, and spin light into tight dense matter. In this EATING OF LIGHT phase, the black hole goes from totally naked empty to finally full. A bright halo occurs as light streams inwards.

In the EATING OF LIGHT phase, the building of a cosmic sphere always begins with a naked black hole as was introduced into the frozen ocean of light at the cosmic origin; or a naked black hole surviving after a star goes Supernova.

If the black hole eats from the frozen ocean of light, this widens the space cavern. Or, the black hole might access light traveling through space.

During an Eating of Light phase, a superspinning naked black hole (left) attracts light, making the light coil and spin into atoms of matter inside of and around the black hole. Spun matter fills the black hole's helix and then adds billions of layers to the body encompassing the black hole.

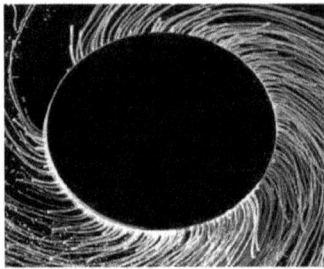

During the Eating of Light phase, light rays are slowed by the gravity of a naked black hole, then curved like a rainbow, and finally spun acutely inwards into the shell of an atom to form matter.

TT Cosmic Rule: The larger a black hole, the more it overeats and the looser it spins atoms while it builds a larger orb, which also means stronger gravity for holding more smaller spheres in orbit.

TT says, 'Cosmic evolution has shown how the largest black holes win out. The winners of the Light Eating battle scene are the gluttonous XL and larger black holes; the competition losers are the small black holes. Cosmic rules dictate, size wins. However, as we will see, overeating is the one great Achilles Heel of black holes.'

TT says, 'At the start of our new solar system, the 177 survival naked black holes engaged in battle for light to build into matter to fill their cores and peripheral bodies. They dueled via their gravity for the best position within the new solar system. TT will show: why a central sun; why solar system has linear shape; why planets revolve around a sun; why moons revolve around planets; why planets have axial tilts and rotate on their axis.'

Phase 2: Lockup of Matter inside of a Sphere

This control of matter is a longest phase. It can last up to 12-14 billion years after the black hole completed eating and spinning light into matter.

Duties for the full black holes switch to the new task of control (LOCKUP), trying hard to prevent atoms of matter from unraveling back to light. Gravity holds things on the surface, plus extends beyond to hold other smaller spheres in orbit.

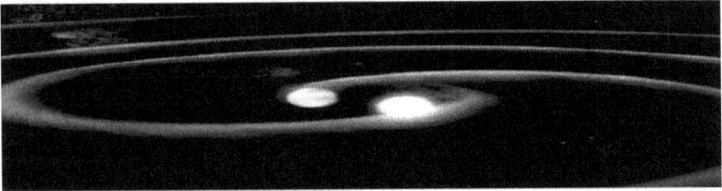

In Lockup, the focus for a black hole changes. Its new jobs are to hold onto matter, plus deal with all other spheres in close proximity to itself.

TT Cosmic Rule: A larger black hole spun looser atoms. It is less proficient at Lockup. E.g. Our Sun.

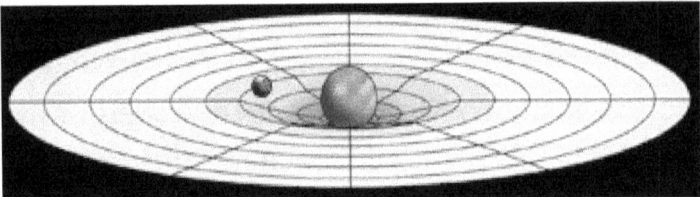

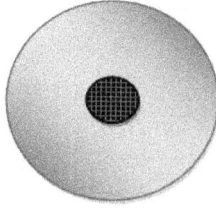

Lockup of Matter phase blends with the earlier Eating of Light phase. Locking up spun atoms is vital, as is the need to entice light until full.

Eventually, the black hole will have spun enough light into matter to be a sphere such as a planet or a moon. The black hole will be hidden, cloaked inside of its sphere. (XL black hole sphere is never seen immediately as a bright sun. First it forms an orb with loosely spun atoms, and then after years of greater loosening it evolves into a sun).

Size Matters: TT distinguishes that the size of a sphere in a solar system is directly related to size of the naked black hole inside the sphere. Once built, the sphere's gravity strength depends upon the size of its black hole plus the sphere's mass. **Distinguish between normal black holes and XL:** The size of a black hole inside a planet or moon can vary from extra small to large. Whereas, the black hole at the center of a sun is an **XL** size.

XL is different: A sun's cycle is different than a planet's or moon's. Right from the day where naked black holes fight for light, an XL black hole battles against all of the other lesser sized black holes. In our solar system, this battle was 1 XL black hole versus 166 lesser black holes.

XL gained an upper hand, eating light sooner and in greater amounts than all the others.

XL spun the loosest atoms in great amounts.

XL's atoms felt the combined tugs of gravity from all the other 166 black holes in orbits.

XL atoms unraveled early in the solar system.

XL sun lost tons of matter back to light.

TT says, 'We might feel sorry for the XL black hole. But, we shouldn't, for its immediate goal is to control an entire solar system. STRATEGY of an XL black hole is to always be many steps ahead. The sun of our solar system will be the first orb in our solar system to shed all of its matter, thus XL will have the upper hand to be an even larger sun in its next solar system.

TT Cosmic Rule: STRATEGY: A black hole never gives away, from one cycle to the next, an advantage it has built up and owns.

A solar system is firstly a battle between naked black holes that build it. Thereafter, cleverly a solar system is a co-operative harmony between cloaked black holes inside of spheres.

The cloaked black holes inside of spheres abide by gravity rules set out to neatly provide harmony to the solar system. Gravity from cloaked black holes provides law and order to a solar system community, with the sun being the town's sheriff.

In our solar system, Lockup certainly is the case with all of the 176 planets and moons. However, our sun is losing its battle at control and lockup, as our sun is into the next phase of Loosening.

So, let's examine our solar system more closely.

Why such a size differential in the size of orbs?

Why the flat disc-like shape to our solar system?

What makes our solar system oddly peculiar?

But first, why isn't our entire cosmos just 1 huge solar system or gigantic galaxy, why billions?

T Theory says, 'Because there are trillions of black holes in cosmos. When naked, these black holes are all battlers, with no love for sharing, coupled with a huge desire to overpower, out-maneuver, and outwit all other naked black holes. They all dream to someday have their own solar system.'

TT says, 'Our solar system is unique. At the end of the old solar system that occupied this Milky Way area, back in the 66th Cosmic Cycle, a Supernova sun Obliterated an old solar system. The XL of the old sun survived and split into two XL naked black holes, as did the smaller black holes inside other spheres. Of an estimated 300 black holes, 124 followed an XL receding to another quadrant; 176 stay with the XL destined as our sun in Cycle 67.

XL ate the fastest to gain greatest mass. Using superior gravity, it gained a hub position. Planets played 2nd fiddle to our XL; moons played 3rd.

But, in out-foxing all of the smaller black holes, XL spun matter too quickly. Matter 'came in hot' in volumes which forfeited any chance of cooling.

In contrast, the planets and moons spun light to matter slowly. Each layer had more time to feel space's coldness. Only tiny centers remained hot.'

T Theory says, 'As we learn more about cosmos, it becomes crystal clear that Formations is ultra important, and that gravity enforces formations. The gravity of a sun controls sphere positioning in a solar system. The gravity of a supermassive black hole controls the positions of all the suns and solar systems within a huge galaxy.'

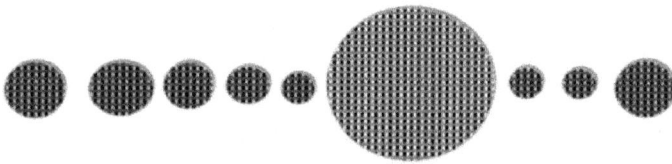

In our solar system, the XL black hole inside the largest sphere (sun) sent gravity out along its equator. All of the 8 planets found orbits within the huge diameter of that pinwheel flat plane.

Orbital direction of our 8 planets follows Cosmic Law stating that planets held in the gravity of a XL black hole all revolve in a similar direction which is determined by the direction of the XL's spin. The XL black hole at the center of our would be sun spun counterclockwise on its axis, so the 8 planets orbited counterclockwise around the XL.

At distances further away from the XL black hole centering our solar system, larger planets gained proximity superiority in their area enabling them to trump the XL's gravity. That allowed Jupiter to hold 63 moons in orbit in its specific gravity field.

Seen all across our cosmos, is this delegation of gravity to lesser sized black holes by the XL black holes of suns and by supermassives of galaxies.

TT says, 'Sphere arrangement in our solar system was decided by the gravity's of the larger black holes. At the hub of our solar system is the XL black hole at the center of our sun, and the next largest black holes inside of planets hold all the lesser sized black holes inside of moons in orbit.'

Rules of Engagement Between Black Holes set the guidelines which organized our solar system. Listed are the 8 planets, closest to furthest from our sun, with diameters and moon numbers.

Mercury (3,032 miles in diameter)(no moons)
Venus (7,521 miles in diameter)(no moons)
Earth (7,926 miles in diameter)(1 moon)
Mars (4,222 miles in diameter)(2 moons)
Jupiter (88,846 miles in diameter) (63 moons)
Saturn (74,898 miles in diameter)(62 moons)
Uranus (31,763 miles in diameter)(27 moons)
Neptune (29,700 miles in diameter)(13 moons)

Totals: 177 orbs which include 1 sun, 8 planets, 168 moons. Cosmic Laws determine that smaller spheres are more easily pulled close to our sun; while most larger orbs stay further away.

Now that our solar system is formed, all of the cloaked black holes are full and hidden inside a planet moon, or our sun. For the rest of their life, the planets/moons will be dedicated to Holding gravity, trying their utmost for strong Lockup of their matter. That effort pays dividends for eons.

TT examines how well these orbs are doing.

Lockup is presently the phase which Earth and its neighbor spheres are in. Earth is doing quite well with Lockup of its matter. Like all orbs, Earth has seen some unraveling of atoms deep inside its black hole. Luckily, Earth hasn't exploded from this expanding pressure, but rather it has utilized volcanic relief. The small to medium sized black hole inside Earth has already had its big Pimple Burst volcano and is now doing a very nice job of relieving pressure via free flowing lava and a few infrequent smaller type volcanoes.

All planets and moons in our solar system likely at some point experienced hot core volcanoes.

TT Says, 'Because of the great cratering on all the moons and planets, our solar system most likely had a past planet which exploded sending rocks into space as meteors crashing into surfaces. Or, some of the Pimple Burst volcanoes had so much power that they spewed tons of rock into space.'

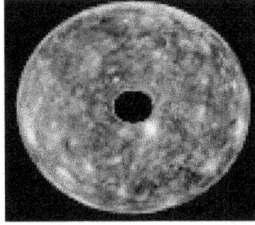

Lockup is presently happening on Mercury. Its smallish cloaked black was easily pulled close to the XL black hole of the sun, which forbade Mercury from having any moons. Even with its smallest size, Mercury has shaken from explosive volcanoes indicating it too has an internal fire.

Saturn presents the most interesting planet with rings circling around its equatorial gravity field.

TT, 'Volcanoes have taken place on nearly every sphere. Saturn's moon Titan has an ice volcano.'

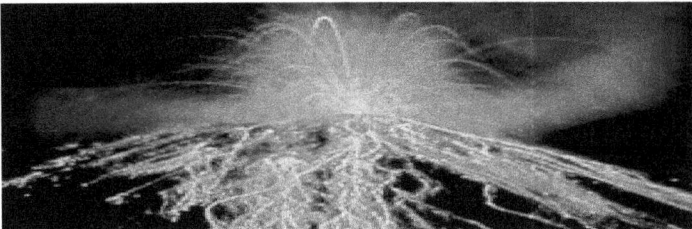

Phase 3: Loosening of Matter inside a Sphere
This lengthy phase begins inside a sphere's core even when it is still in a lengthy Holding phase.

T Theory says, 'Matter always unravels back to light, even if it takes billions of years. The gravity of a black hole can only keep matter in Lockup for so long. The upper limit for a solid sphere is 15 billion years. Whereas, eventual LOOSENING and ESCAPE of matter occurs earlier with a sun.'

The following illustrations show the series of events which take place inside of a black hole and its sphere as atoms of hard matter begin to loosen and unspin and turn into hot lava.

The nuclei of the oldest atoms begin to loosen as they spin less tightly. They expand and unspin in their attempt at escape. But, there's nowhere to go for loosened atoms at the bowels of a black hole deep inside a sphere, buried below miles of solid matter. Yet, loosening still creates heat and expansion. A fire percolates deep inside.

The black hole centering a planet will eventually lose control of its oldest spun light. Deep down inside of its bowels, matter unravels into a fire.

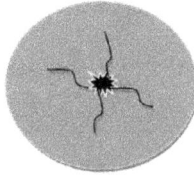

As more atoms loosen in the core of the black hole inside of the sphere, the interior fire grows. As internal pressure mounts, the sphere swells. Fissures form, searching for a route to the surface.

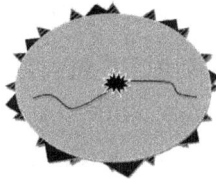

From the lava's pressure, fissures buckle surface areas. Gigantic earthquakes cause upheaval as mountains of rock are forced to the surface.

Over the following billion years or so, the extent of the fire inside the planet will depend on size. The smaller the planet or moon, the smaller is its the black hole and the tighter the matter. Therein, moons and small planets will experience a slower gradual unspinning of all their core matter, and a slower increase in the size of the internal fire.

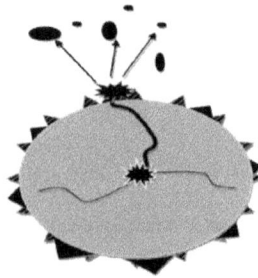

The fire, which started deep in the belly of the black hole, is finally ready to break through to the surface of the planet. The pressure for the planet to expand is so great that the fire must reach the surface to prevent a sphere explosion. Finally, a breakthrough occurs, with the historical inaugural volcano known as Pimple Burst. This volcano is so enormously powerful, that pent-up rock is fired like a rocket out into space. For many millions of year, these meteorite rocks can collide into other moons and planets in the vicinity, cratering them.

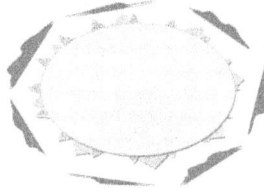

After a rash of volcanoes, the sphere experiences a cold Ice Age as ash cloud blocks out light entry.

After the Ice Age, the sphere's surface re-warms. While volcanoes still occur, they are less violent. More fissures reach the surface as lava flows. However, the continual internal fire causes orb expansion with continental drift prevalent along the entire outer surface of the sphere.

For a black hole inside of a sphere, to control all of this matter is a long tedious job. Light always Loosens, always eventually escaping from its atom, even if it takes billions of years.

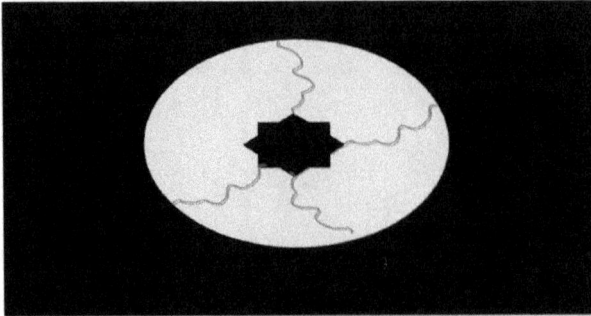

The fire inside the black hole intensifies. At times, observance of any internal fire may subside, yet home fires still burn, and the size of the fire in the internal core increases. It is only fissures leading to the surface that help relieve sphere pressure.

For the larger planets, such as Jupiter in our solar system, it really saw its black hole overeat during construction, so its interior is extremely hot. This heat has expanded drastically and thus turned the planet's surface into a gaseous liquid state.

TT studies the loosening which caused our sun. It is relevant to recall that the XL black hole which formed our sun spun atoms a lot earlier and a lot looser than the all other lesser sized black holes.

When our solar system was forming, there was only one XL BLACK HOLE out of the 177. The XL ate by far the most light, getting the jump on all the lesser sized black holes. The gluttonous XL was a voracious eater of light, the Henry VIII of our solar system, eating until ready to bust. The XL black hole grew into a gigantic loose orb.

The XL black hole also got central sun status as its dominant gravity force allowed it to hold all lesser sized spheres (planets/moons) in orbits.

The control freak XL kept all the smaller orbs in orbits. Ironically, the powerhouse XL was the first to loosen. The kryptonite of a XL black hole is that it always overeats, devouring light too quickly.

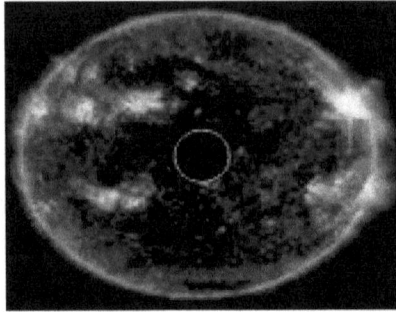

With the job of holding a host of other spheres in orbit, the gravity from all of those other black holes pulled hard on the contents of XL's matter.

The XL black hole inside of the gigantic sphere easily lost more control of its matter. Its surface became liquid as the interior lava encroached upwards. The orb turned totally hot, conquering even the perpetual cold of space. As an inside fire intensified, the hot furnace began to evolve into a sun. Hot atoms unspun on the surface, escaping the orb as light, free to travel the space highway.

Because of the expansion of matter, and because of the amount of matter which a black hole can build, even the XL black hole at the core of the planet-turned-sun was small in comparison to the body of the sun it created around itself.

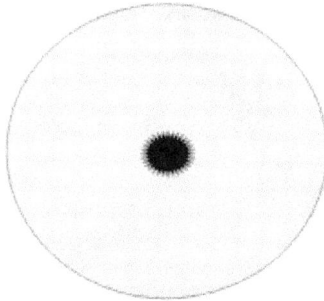

Extraordinary is the intense heat of the XL sun, as unspinning atoms see surface temperatures over 5,000 degrees. Solar flares can erupt hundreds of miles out into space. A rapid succession of huge solar flares may be the omen that an XL sun is preparing to go Supernova and wipe out its solar system. For the moment, Earth appears safe.

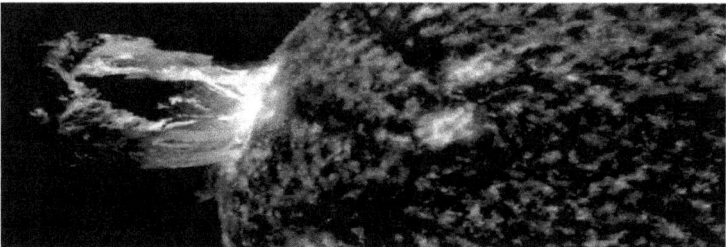

Phase 4: Supernova of a Sun, and escape of light, during the Obliteration of a Solar System.
This phase happens to a sun and its solar system when reaching the 10-15 billion year age range. This phase is abrupt, as Supernovae occur quickly as does the follow-up solar system Obliteration.

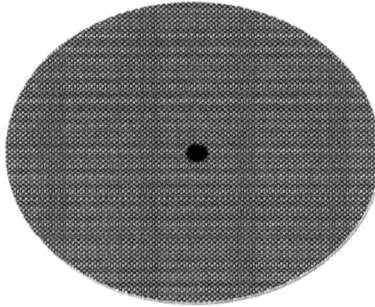

Phase four deals with the final episodes in the life cycle of a sun. Very few spheres reach sun status, as those of lesser size are destroyed within their solar system when their sun goes Supernova.

Just how late in its cycle a sun goes Supernova is certainly up for debate. Certainly, as a sun ages, the XL black hole deep inside steadily loses more control over its matter. However, a big factor as to how long a sun lasts may be external, such as the accidental appearance of a naked black hole which would quickly and totally unravel a sun.

The image below shows that as a sun ages, the sun further expands as the XL black hole at the sun's core loses more control over its atoms.

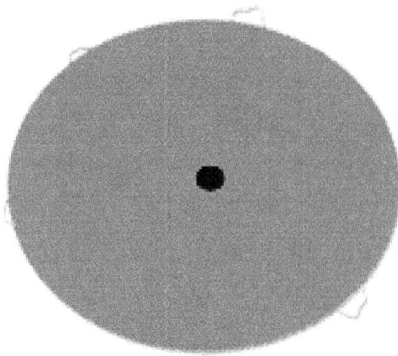

The sun continues to grow larger and hotter. The X-large black hole at its core can barely keep control. Luckily, the black hole's gravity keeps the release of atoms on the sun's surface from being totally explosive, and the large mass of the sun is still able to keep its entire solar system in orbit. But, with the passage of time, this sphere is now 12 billion years old. Each day more light breaks free after billions of years in Lockup as an atom. For now, the solar system is nicely heated. But, if this sun overheats and its solar flares become too gigantic, that signals that this sun is near its end and that spells a coming end to the solar system.

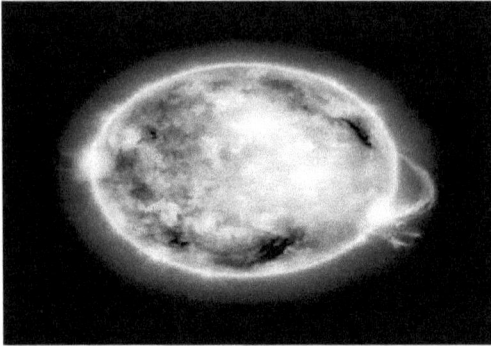

The elderly sun, now over 13 billion years old, expands into a colossal mammoth old star, ready to die. At the sun's surface, photosphere helium atoms break apart as light is emitted to freedom after many billions of years of captivity. Finally, the dying star loses entire control, exploding into the brightest cosmic sight, a Supernova. If there happens to be a rogue naked black hole close to the Supernova, a bright Quasar is seen as a rush of light from the sun goes into the predator.

Supernova. The image above if of a naked black feeding by eating steams of light from an old sun which is going Supernova. The black hole eater of light is building a new sphere around itself. Now, this naked black hole may have just happened to be in the area of a sun, wherein the naked black hole hastened the Supernova death of the star.

TT proposes yet one other possibility. Take a scenario where the Supernova of a sun already occurred, and the black hole is the survivor from the core of that sun. The XL black hole is eating light from a large exploding planet which was hastened into a Supernova when all the planets and moons of the solar system were Obliterated by the Supernova. Therein, the black hole survivor from the exploded sun is the first black hole to begin eating light to build a new solar system.

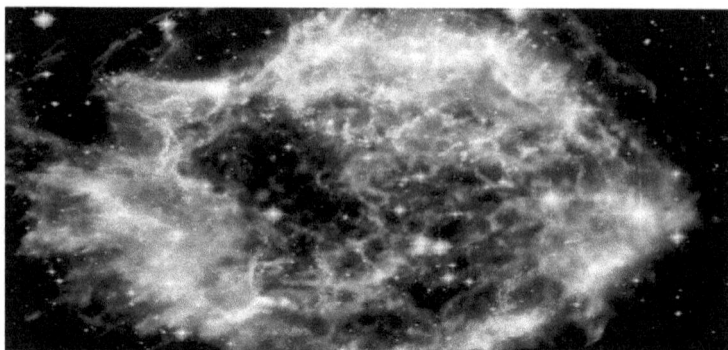

The end of a old tired sun's life is its explosion as a Supernova. These Supernovae occur each day in our cosmos, where a sun explodes, wiping out (Obliterating) a solar system. Seems mean. Why can't a sun just die quietly. Why the explosion?

  Here is what TT says, 'A Supernova is more than just a show put on by a sun. The XL black hole resident at the core of the sun has a STRATEGY. It needs to be the sun of the next solar system as well. By quickly getting naked via a Supernova, a XL can get the upper hand on all the other black holes inside of the planets and moons of its old solar system. XL gobbles up the Supernova light (seen as an implosion back into the black hole). XL also consumes light from all the lesser sized spheres which incur their own explosions during the solar system Obliteration. This guarantees in the next cycle that XL is will get even mightier.'

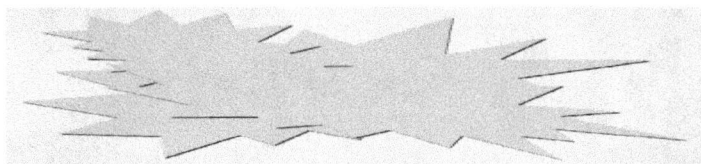

An interesting query here is: to what extent does a sun destroy its own solar system when it goes Supernova? T Theory calls this event Obliteration. In our solar system there are 8 planets, and 168 moons, plus one sun. If our sun went Supernova, the power burst from that exploded star would likely destroy nearly every planet and moon in our solar system. In an Obliteration, every planet and moon melts and explodes doing its own smaller Supernova. The only orbs spared might be those spheres in distant orbits from the sun.

   T Theory states, 'After a Supernova of a sun, the black hole survives which was at the central core of all the matter of that sun. Furthermore, each black hole at the core of every planet and moon survives the Obliteration of its old solar system.'

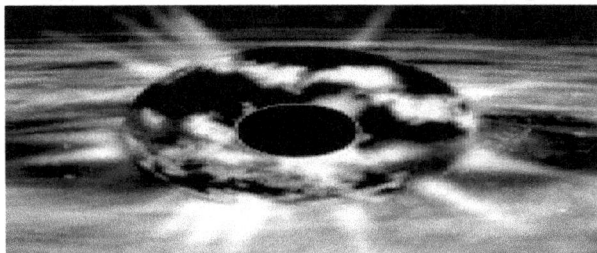

Imagine the blast when any sun goes Supernova. The heat from all the instantaneously escaping light melts all the orbs in that sun's solar system. This Obliteration destroys all the surfaces, leaving all the resident black holes from the core of every sphere as the only survivors – naked survivors.

After Supernova and Obliteration, tons of pent-up light escapes from the matter which had been locked up on and within all those spheres. That freed light departs, free to travel space, or until getting lured inside by a black hole gravitational station, to once again be spun back into matter.

Much of the happy freed light only experiences temporary freedom, as all the now naked black holes which survived Obliteration commence a battle against one another to spin that light into new atoms of matter for the construction of new spheres for the building of a new solar system.

Phase 5: Replication Reproduction of Black Hole.
Short phase occurs after 13-15 billion years, and follows on the heels of Supernova. However, this could also be called year zero, as this is the start of the new solar system which replaces the old.

The pliable elastic naked black hole can never be destroyed, not even by a huge Supernova. The black hole survives a Supernova by splitting into two. Now, two black holes replace the old one.

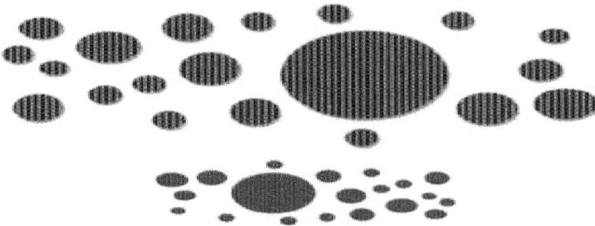

TT **says,** 'After Supernova and the Obliteration of a solar system, there may be dozens to hundreds of naked black holes which were at the center of the old sun, planets, and moons of that destroyed solar system. All the naked black holes comprise a survival graveyard where each will already have split in two (Replicated) from the great force of the Supernova and the Obliteration.'

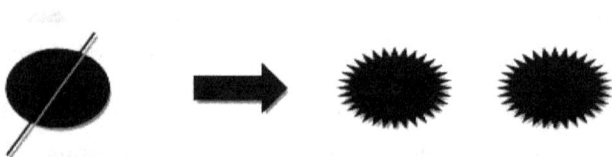

TT can only guess as to how a black hole can split into two. When Supernova happens, the exposed naked black hole instantaneously reverts back to an incredible fast spin rate. The violent Supernova whiplash snaps the spiral helix of the black hole into two, thereby creating two new black holes.

Replication of a black hole is an end phase, but also a beginning phase for a new solar system. **TT says,** 'It's rare for an XL black hole split to result into two equal twins. Black holes are strategists, wanting to be the biggest and most controlling. The XL black hole splits making one top-dog XL, and one smaller to ensure future victory.'

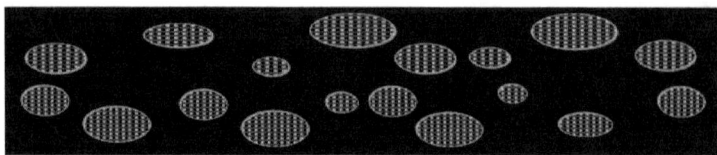

TT says, 'All of the newly naked black holes in the graveyard of a Supernova and Obliteration start out as reborn. But, their DNA (so to speak) may date back 100s of billions of years to past cycles.'

TT says, 'In our solar system, hidden away inside of every sphere is a cloaked black hole. Our solar system exists during the trillionth year of our cosmos. We live in the 67th of the 15 billion year generations of spheres and solar systems inside of the cosmic galaxies. Thus, a black hole at the center of a sphere such as Earth may have ties dating back a trillion years to the cosmic origin.'

Review 5 major phases in a 15 billion year cycle. TT summarizes how our solar system is doing.

EATING LIGHT BY BLACK HOLES (the building of spheres was completed 5-8 billion years ago).

LOCKUP INSIDE OF SPHERES (on going).

LOOSENING OF MATTER (has begun as lava deep inside Earth. Whereas, our sun is loosening fast).

SUPERNOVA AND ESCAPE (in the future).

REPLICATION REPRODUCTION (in the future).

Earth is in Lockup coupled with small Loosening. Sun is Loosening, headed someday to Supernova.

TT does an even closer examination; performing forensics on our current solar system as to orbital directions and axial spin directions. The sole XL black hole is inside of our sun. The 8 planets are in flat plane orbits around the equator of the sun.

TT says, 'Orbital directions of our 8 planets follow the counterclockwise axial spin direction of the XL black hole at the center of our sun. The gravity that our sun projects is counterclockwise, so the 8 planets all orbit counterclockwise around the XL.'

All the 166 moons in our solar system tag along in harmony with their individual planet, orbiting in a counterclockwise direction around our sun. But, moons also have an additional orbit around their own planet. Every moon can orbit clockwise or counterclockwise around its planet, depending on the direction of axial spin of the planet.

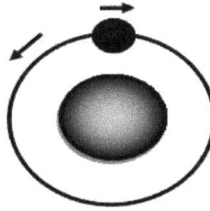

TT says, 'It is important to understand this easy terminology; differentiate revolve and rotate.'
- in the diagram, the left arrow shows that a planet <u>revolves</u> (orbits) around a sun. And, a moon revolves orbits) around a planet.
- in the diagram, the top arrow shows that a sphere <u>rotates </u>(spins, pivots) on its axis.

In a Big Bang or Nebula theory cosmos, there is a supposition that everything should revolve and spin in a direction set out by the swirling nebulae that created spheres, solar systems, and galaxies throughout the cosmos. However, that isn't the case. No one can explain why.
TT shows how revolve and rotate really work.

T Theory explains why a planet (or moon) can revolve (orbit) in one direction, but rotate (spin and pivot) in an opposite direction on its axis.

TT says, 'We live in a solar system where our head master, our sun, rotates (spins) counterclockwise on its axis, thereby extending a counterclockwise gravity to the entire solar system. Does this imply, as Big Bang and Nebular theory contends, that all our planets and moons in our solar system must follow the lead of our sun and rotate (spin, pivot) counterclockwise on their axis. **TT declares , no.'**

TT says, 'With every sphere, there are two main forces at work. Firstly, there is a gravitational pull from a larger sphere with a larger black hole at its core. E.g. Earth is forced to orbit our larger sun. Secondly, each black hole inside of a sphere can independently determine its own axial direction. E.g. Earth rotates (pivots) counterclockwise on its axis, while Venus spins clockwise on its axis.'

 TT states, 'Black holes independently determine their own axial spin direction. Thus, our cosmos is very 50-50 at being both right and left handed.

 But, once a black hole determines which way it will spin on its axis, the only thing that can ever change that is when the black hole splits in two. At that moment, the 2 black holes get their spin direction set for their next 15 billion year cycle.'

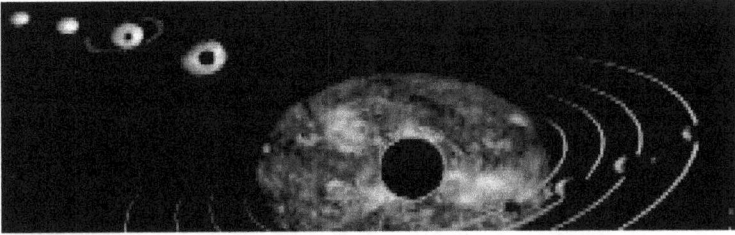

TT is the only theory that solves this puzzle.

T Theory says, 'In our solar system, black holes inside of planets Mercury, Earth, Mars, Jupiter, Saturn, Neptune, all rotate counterclockwise on their axis, as does our sun. But, the black holes of the weirdos Venus and Uranus don't.

Astronomers feel all of the planets should spin on their axis in the same direction as our sun. They can't explain why Venus and Uranus differ.

TT answers this mystery, 'The direction of axial spin for a planet (or moon) is independent of the sun's spin direction. In our solar system, 6 planets spin counter and 2 spin clockwise. So, 6 planets (and our sun) have black holes inside which spin counterclockwise. While the weirdos Venus and Uranus have clockwise black holes.

TT does however admit that the spin direction of a XL black hole via its powerful gravity draft might be able to slightly influence the selection of spin direction for smaller black holes.'

Next, TT takes a look at the cosmic hotels, the ancient galaxies which define our cosmos.

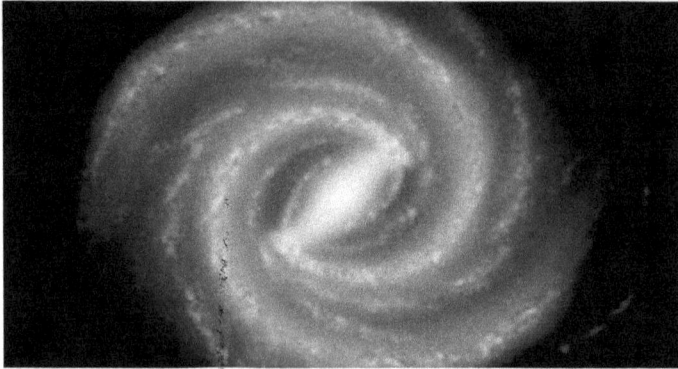

These ancient galaxies are in the rental business. They lease spots in their galaxy to solar systems. A top restriction for a rental lease within a galaxy is that the solar system's sun must go Supernova some time prior to 15 billion years. In this way, solar systems die and are reborn larger and in greater numbers, thereby guaranteeing eternal growth for the humongous ancient galaxy.

T **Theory says,** 'A solar system's sun is but one of billions of suns with solar systems residing within an ancient galaxy. A sun without a solar system is a rare anomaly. These galaxies have at their hub an ancient supermassive, 500-800 billion year old, black hole in control of the galaxy. With each new 15 billion year era, the solar systems of the galaxy take turns recycling after Supernova of their sun. Each cycle, solar systems grow more numerous as many larger solar systems split apart forming two new solar systems. Over time, the number of stars and solar systems within the ancient galaxy grow. The galaxy is destined to grow forever larger.'

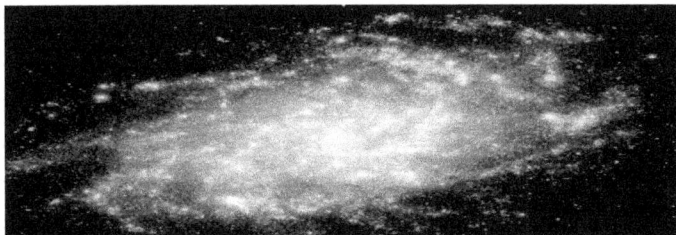

Do galaxies spin clockwise or counterclockwise?
Astronomers find this to be a 50-50 proposition, which definitely supports TT which says, 'Cosmos is both left and right handed, meaning a galaxy can for its life time spin either counterclockwise or clockwise as determined by the supermassive black hole at the hub of the galaxy. When the supermassive first formed a galaxy, as long ago as 800 billion years, the axial spin that was set at that time inside the supermassive, has been and will remain with that galaxy forever.

All of the stars, suns, and solar systems within the body and spiral arms of a spiral galaxy will orbit (revolve) around the supermassive as set by the gravity direction extended outwards from the supermassive as it spins on it axis.

Independent of the supermassive black hole at the hub, the sun of every solar system within the galaxy freely sets its own axial spin direction.

Spiral galaxies are definitely the most ominous sights in our cosmos. They present themselves with power, majesty, and magnificence as they show their splendor and flaunt their spectacular beauty. Galaxies are wondrous yet solitary. They can be described as gigantic remote private islands in space. Our present cosmos holds over two hundred billion galaxies, each containing millions of stars with solar systems.

While solar systems recycle on average every 15 billion years, galaxies are permanent since they don't live and die by the 15 billion year rule. Solar systems recycle within galaxies each 15 billion years thereby fostering growth so galaxies can get ever larger and become more populous each time a solar systems recycles.

T Theory portrays a whole new picture as to how galaxies came to be and how they grew so large.

TT says, 'Think of a galaxy as an island cosmic hotel. The galaxy has millions of hotel guests, each residing as a solar system. No matter how many times the solar systems within the hotel check out of rooms, the number of total rooms always increases. Occupancy rate grows, yet the galaxy hotel remains the main entity.

A gigantic galaxy easily survives a Supernova recycle of one of its solar systems while the rest of the galaxy is unaffected. While all of its solar systems get reborn every 15 billion years, the galaxy is likely well over a half trillion years old.

The oldest largest galaxies have gone through as many as 30-50 of the 15 billion year recycles of their solar systems over the past 600-800 billion years. Estimate the number of stars in a galaxy and it is possible to closely calculate the galaxy's age by using an exponential doubling premise each 15 billion years. Galaxies are by far the best recorders of cosmic history, having witnessed hundreds of billions of years.'

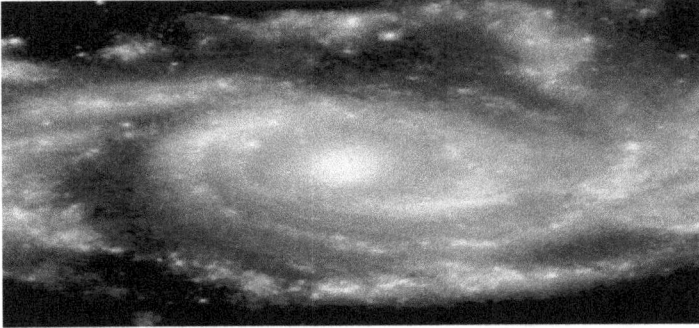

Spirals make up about 90% of our universe's galaxies. Their stars and solar systems are in orbit around a galactic bulging center.

This bulge has enormous mass with enough gravitational power to hold in orbit millions of stars with solar systems. At the very core of the bulge is a fast spinning supermassive black hole surrounded by a clustered of superstars.

T Theory says, 'That bulge is where Cosmic Laws of Supermassive Black Holes are rewritten.'

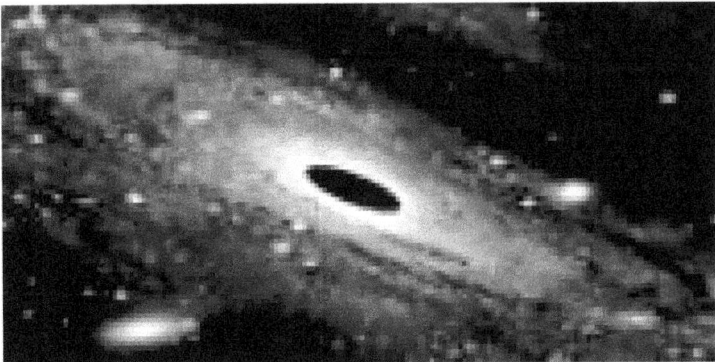

T Theory talks galaxy evolution, 'Starting out, the first galaxy began nearly a trillion years ago. Black holes were the galaxy builders. Naked black holes ate from the frozen ocean of light. The interaction of these black holes formed the first solar system. This first system grew larger when its star went Supernova, doubling the total number of black holes inside of spheres. Recycles took place every 15 billion years, making for more solar systems.

The first galaxy was formed when several solar systems came under orbital control of the central massive star governing the largest solar system. Later, that small galaxy underwent a Supernova splitting the massive black hole apart to form two galaxies. Thus, a doubling process provided more galaxies. Today, billions of galaxies house billions of solar systems. Yet, it was the evolution of a massive black hole to a supermassive black hole status as the chief reason for gigantic galaxies.'

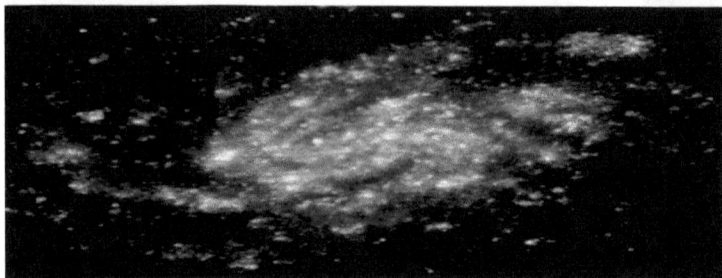

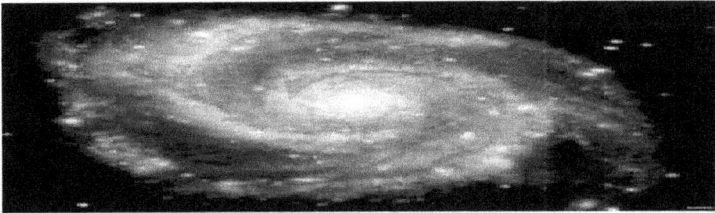

So, let's say that a giant galaxy is ¾ the way to being a trillion years old. How did a galaxy grow that old and big, and why is the black hole at the hub of the galaxy so supermassive.

T Theory says, 'In cosmos, size matters, referring to black holes and their powerful gravity. Eg: The largeness of planet Jupiter's black hole allows it to hold 63 moons in orbit; the XL black hole of our sun holds 176 orbs in orbits. Massive suns, many times our sun's size, have massive black holes controlling even larger solar systems. Then, in mammoth galaxies, the largest existing black holes, supermassives at the hub of galaxies, hold millions of solar systems in orbit.'

'The question becomes: were black holes given different sizing at their origin, or did they deploy a means of growing larger over eons of time?'

TT says, 'Think of cosmic evolution as survival of the fittest (best adapter). So, when thinking of the evolution of black holes, their goal is to control as much matter and as many orbs as possible. The strategy deployed by black holes to fulfill these goals is to grow ever larger. Thus, they utilize their size advantage to out-duel lesser sized black holes in a battle for light so they can grow ever larger. They exploit their own elasticity enabling themselves to expand their size as they gluttonously over eat.

The prolific masters of all of these tremendous feats are the ancient supermassive black holes found at the hubs of spiral galaxies.'

The supermassive black holes at the each hub of galaxies evolved, able to grow ever larger. As TT will show that they adapted their inner structures, ensuring their victory.

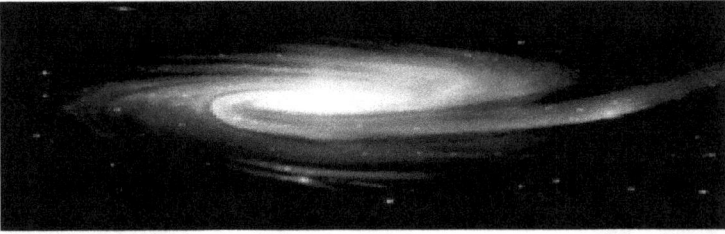

Massive versus supermassive black hole.

The problem for a massive black hole at the hub of a small galaxy was with the Cosmic Laws. Now that the black hole had grown massive, and as a massive sun controlled many solar systems in the form of a small galaxy, it wanted to maintain this dominance from one 15 billion year cycle to the next. But, Cosmic Law said 'no.' Near the end of its 15 billion years, as a massive sun, Cosmic Law made it go Supernova and Obliteration blew the small galaxy apart. Sure all this meant more small galaxies, but the massive black hole hated that its galaxy was unsustainable for a longer period and always had to be rebuilt – the massive black hole had to evolve by adapting a better way.

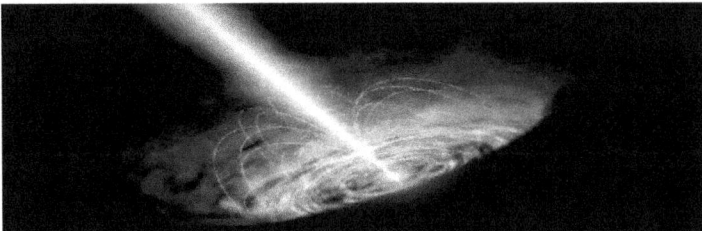

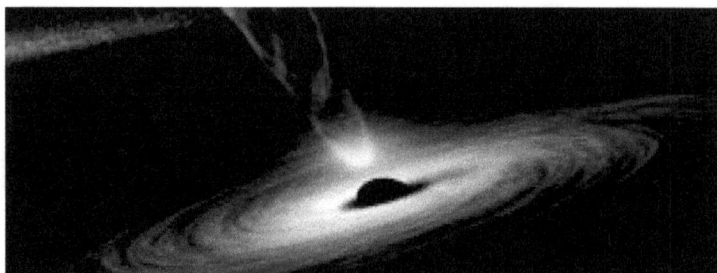

When massive black holes reach a certain bench mark size, they are able to evolve by the way they adapt to the eating of light. While still pulling tons of light towards them and devouring it, they have changed what they do with this light.

The massive black hole, becoming supermassive, determined not to eat light solely for the building of a supermassive sun around itself – that was too unproductive. It found that building a sun meant eventually having to go Supernova detrimentally destroying its galaxy. A smarter strategy allowed the supermassive black hole to evolve to be much older and control its galaxy for longer than just one 15 billion year cycle. Instead, supermassive's galaxy could grow for 100's of billions of years. This strategy allowed the supermassive black hole to become a true monarch, building an entire empire of suns and solar systems inside its galaxy.

The new strategy for a massive black hole at the hub of its very young galaxy was to avoid having to build a sun around itself and eventually have to go Supernova. Adaptation for a massive black hole meant making structural changes allowing it to deal in a new way with the tons of light which it pulled inwards. Change allowed a massive hole to keep its position at the hub of its galaxy, while the massive black hole grew to supermassive size.

If TT is correct, and a black hole can adapt its structures in evolving over hundreds of billions of years, that means we may have no choice but to view black holes as living entities. But, they are so different than all of the living things we know of. Black holes are indestructible and supermassive black holes can live longer than a trillion years.

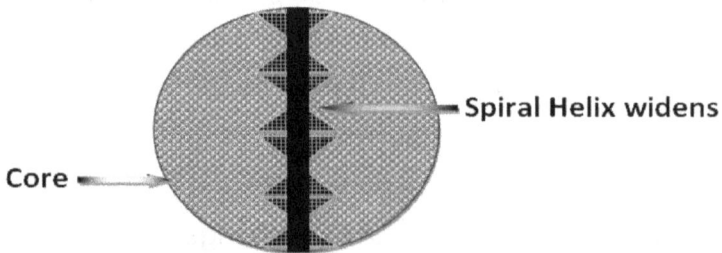

Core ⟶

Spiral Helix widens ⟵

The supermassive black hole had to make new structural changes in order to evolve. Changes involved a widening of its spiral helix allowing it to become more hollow. As it ingested tons of light, it only spun enough light to matter to continually fill its elastic compartments, thereby growing in size, without having to build a sun. The wide hollow helix became a conduit to pass digested light out to the exits. In essence, a breathing ability was developed. Supermassive black holes display this as plumes of light jettisoning from the top and bottom of the supermassive black hole which pumps the light. Thus, a supermassive strategized allowing it to live indefinitely at the hub of a galaxy.

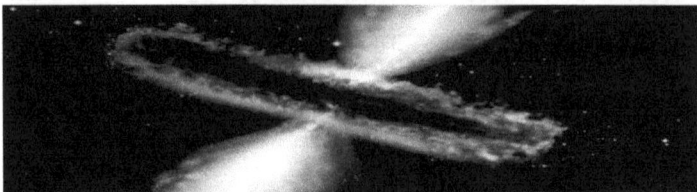

T Theory says, "Some supermassive black holes have been at the center of their galaxy for 800 billion years, and destined to remain in that dominant position for ever. Changes in function have allow supermassive to win on many fronts.

Firstly, the supermassive doesn't have to grow into a sun. It can live long to control its galaxy.

2nd, it can spin fast because it stays as a black hole, never taking on a sun's sluggish rotation.

3rd, it can pull huge stars into clusters around itself, having to ingest their light, yet it can spew and jettison that light back out as plumes.

4th, it can use the mass of the stars which it pulls into clusters around itself to assist with the immense gravity needed to hold the entire galaxy together including the spiral arms.

So for supermassive, humongous makes a big difference. The supermassive black hole can live on through hundreds of billion of years as the stars and solar systems recycle over and over inside of its galaxy. The rich get ever richer.'

Clearly seeing a galaxy's supermassive black hole is not easy since a bright halo from the massive surrounding suns blurs out our view of the supermassive black hole. T Theory says, 'Strategy by the supermassive black hole allows it to pull dominant stars into a cluster around itself. Those stars get gobbled up and never permitted a chance to challenge for supremacy of the galaxy. Although, the mass of the stars assist with the gravitational pull necessary to hold the gigantic galaxy together. With this ingenious strategy, the supermassive black hole uses its prowess to make the stars subservient in aiding to control the humongous galaxy.

The pulled-in stars glow brilliantly as there is a whole plethora of these stars in concentration surrounding the supermassive black hole.

There is always more and more light needed by a supermassive black hole to run the operations of a galaxy. Thus, galaxies are continually attracted outwards towards the perimeter of our cosmos to be closer to the available ocean of frozen light. At this perimeter, many new worker-bee black holes are busy breaking off chunks of the static light, trying to consume it. But, uncaptured light travels through space becoming available to the current galaxies. Therefore, galaxies recede outwards as they try to get closer to this well of light coming from the outer perimeter of created space.

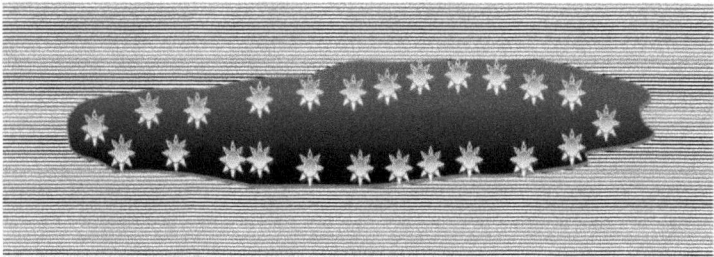

TT says, 'It is folly to attempt to fit the ancient age of a supermassive black hole into an astronomer's 13.7 billion year estimate for our cosmos. Once TT proves the age of supermassive black holes at 200 billion to 800 billion years, the new estimate of the origin of cosmos, and its age, will escalate immediately upwards to TT's TRILLION years.'

Evolution of a black hole to supermassive status, centering a galaxy, took 100's of billions of years. **TT says,** 'The story of these ancients promises to be the longest saga ever told.'

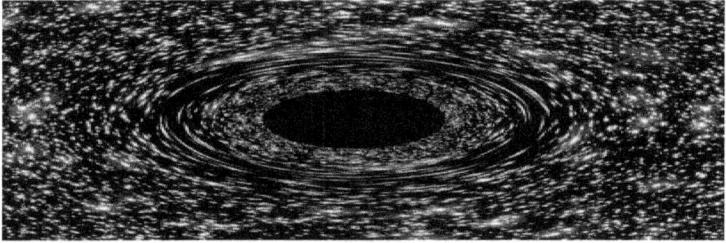

The above image shows an artistic depiction of a supermassive black hole. Astronomers, In April 2016, uncovered a record breaking supermassive black hole in an area which is sparsely populated. Its weight was equivalent to 17 billion suns. The discovery suggests that supermassives may be more common than previously thought.

TT says, 'Supermassive black holes exist at the center of galaxies all across the cosmos. A galaxy without a supermassive at its hub really can't exist, since supermassive black holes were (are) builders of the billions of cosmic galaxies over the past trillion years of cosmic history.'

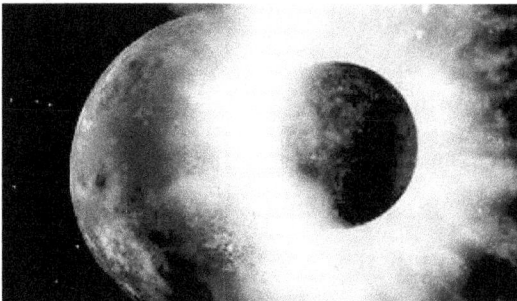

TT says, 'However, don't let size of a black hole fool you as to its age. Ancient supermassives can easily show their age of 800 billion years. But, a small black hole inside of a planet or moon may also be ancient, having resided and been inside a new sphere during each 15 billion year cycle. With its small size, it lost all the battles to larger black holes and thus has simply stayed small over a trillion years of cosmic history.'

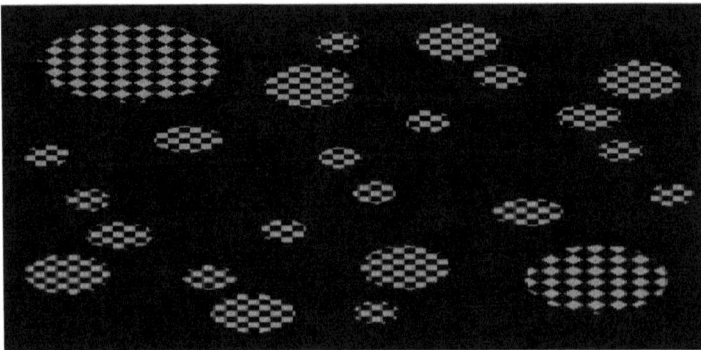

Age of our Cosmos.

T Theory says, 'Yes, our cosmos is a trillion years ancient. The discovery of the presence of an area of black holes following a Supernova basically would prove and substantiate T Theory's claim that our cosmos recycled and grew over a history far older than just a youngish 13.7 billion years.'

Each one of the historic 67 cycles of our cosmos lasted approximately 15 billion years. There is no exact cut off from one cycle to the next, rather one cycle overlaps with ties into the next cycle.

Take our solar system, we're about 1/2 the way through our present 15 billion year cycle of our sun. After our sun goes supernova, another 15 billion year solar system will start up from the surviving black holes which are at the core of every sphere in our present solar system. Yet, all our neighboring solar systems might be younger or older, on a totally different recycling schedule.

T Theory states that cosmos is a trillion years old. Each of the 67 bars represents just 1 of the 15 billion year eras that took place during a trillion years of cosmic history.

T Theory says, 'Can't see cosmos for the stars. Ancient cosmic history is hidden away by the recycling process of spheres and solar systems. People incorrectly miscalculate our universe's age using the oldest present-living stars, which are about 13.7 billion years old, as their gauge. Unfortunately, cosmic recycling blurred away all traces to the past. Each of the past 67 cycles, each about 15 billion years in duration, is lost into history. While, our new 15 billion year cycle appears totally unique and provides the illusion of being the one and only solitary cycle.'

TT says, 'A galaxy may appear to be only 13.7 billion years old, as that may be the age of the oldest stars and solar systems.

But, look deeper past the solar systems to the supermassive black hole at the galaxy hub, to a black hole which is over 800 billion years old.'

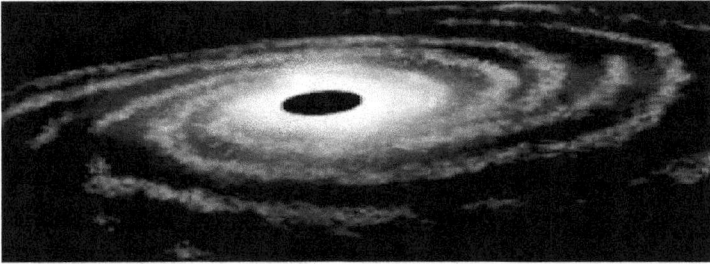

In an 800 billion year old galaxy, the stars and solar systems are less than 15 billion years.

Like a hotel, over each 15 billion years era, the galaxy changes its occupant solar systems.

TT says, 'Let's count the spheres in cosmos. Astronomers estimate 73 quintillion.

The following mathematical progression shows TT doubling the star population of cosmos each 15 billion years. Exponentially, the numbers start as just 1 star in cycle 1, doubling with each and every new cycle, then dramatically escalating.

Cycle 65 shows 18,446,744,073,709,551,616 stars.

Cycle 66 shows 36,893,488,147,419,103,232 stars.

Cycle 67 shows 73,786,976,294,838,206,464 stars.

What started tiny, is now 73 quintillion stars in the 67th cycle of our cosmos in its trillionth year. We reside in the latest 15 billion year cycle.

15 billion X 67 recycles = 1,000,050,000,000 years.

TT says, 'But, there's even more. Suppose TT is correct in stating that a black hole resides inside of every sphere. So, redo the math with about 100 spheres per solar system. That means that 100 X 7,300,000,000,000,000,000,000 does equate to 7.3 sextillion orbs with black holes in cosmos.'

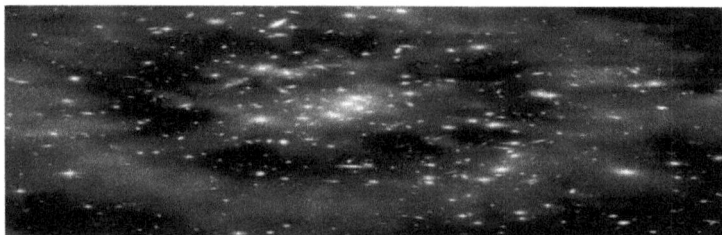

A concession: TT's trillion years since inception, is an based upon how cosmos is growing. However, it is conceded that the initial onset of our cosmos may have begun with an injection of millions of black holes as a means of super-speeding cosmic growth. If so, cosmos could be a little younger than a trillion, at only 900 billion years or so.

In the next 15 billion year cosmic cycle (68), size will again double. Within galaxies, solar systems will double in number. The total cosmic star count will double from 73 quintillion stars in our present 67th cycle to 146 quintillion stars in the 68th cycle. The greatest growth era in the past trillion years.

TT says, 'The cosmos is trying to give us signs of this fantastic population growth. The brightness of a Supernova is saying, 'Look! See an exploding star; see obliteration of a solar system; see the after affects as new naked black holes fight for light as they begin building a new solar system.'

Time. T Theory's definition of time is far different than proposed by anyone else. It includes: what time is; how it was imported into cosmos; plus where it does or doesn't exist inside cosmos.

A trillion years is an awfully long time, especially for impetuous impatient humans. Yet, for cosmos and its black holes, eons of time is quite ordinary.

T Theory's definition, 'Time is imported inside of our cosmos right along with the spinning of light into matter. Wherever atoms of matter are confined in spin, a clock ticks measuring the duration during which light spins below light speed as it tries to loosen and escape.'

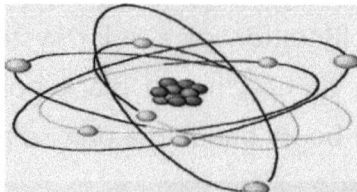

Time is a concept invented specifically for our universe. At the cosmic origin, there was only the static light ocean and time was locked up inside of the zillions of strands of light, waiting.

TT says, 'When light strands escaped from the light ocean, they moved to the full speed of light. At light speed, time stood still inside of light's toolbox – waiting to be activated. '

Alberta Einstein said, 'Time is an illusion.' Einstein showed how time slowed as it neared light speed. At light speed, time stood still. An hour glass on its side shows time standing still.

TT says, 'Time is released from light's genie bottle whenever light moves below light speed. In our cosmos, time pops out when light meets a black hole, either naked or at the center of a sphere.'

Where in cosmos does time exist or not exist?
T Theory says, 'Time doesn't exist in (1) the ocean of frozen light. Time is in the light ocean, but held static in readiness inside the frozen light strands.

Time doesn't exist in (2) empty space or when a light ray travels at its full speed through space.

On a flip side, time does exist in (3) where light's speed is slowed by the gravity of a black hole, or the gravity of a black hole inside of a sphere.

Thus, Earth, with a black hole inside its core, is a zone where light is slowed by Earth's black hole gravity and therefore on Earth time exists. In our solar system, time exists in and around the gravity fields of all the planets, moons, and our sun.'

Let's say that a ray of light travels from another galaxy. That ray travels at the full speed of light through space, so time is locked away. But, as the ray approaches the gravitational station of our Earth, which has a black hole inside, light slows and time is imported. So, a gravity which can slow light's speed is necessary for time to exist.

How was time invented? Time likely doesn't tick outside our universe. That no-time realm, doesn't infer any beginning or end, but a continuance to move forwards and backwards through events.

Time is a specialty item created for our universe. We're a destination oasis for any visitant wanting to experience the peculiarity known as time. An oasis where the cosmic inventors ingeniously tied time to the speed of light.

Time, carried by light, springs from light's tool box, imported into a zone when light's speed is slowed by the gravity of a black hole, or when spun into atoms of matter by a black hole.

Spheroids, such as Earth, experience time where birth, growth and aging are fashionable. But, for any cosmic explorers or caretakers able to travel at light speed, time is nicely tucked away.

Stated again in T Theory, 'Time is defined as a measure of the duration during which light slows below its full speed wherever it encounters the gravitational pull of a naked black hole or the gravity of a black hole cloaked away inside of a planet, moon, star, solar system, or galaxy. Visit a black hole sphere to experience time. Ride a light ray through empty space to experience no-time.'

It took 5 steps for time to be a universe feature:

Step 1: Time was scientifically invented, but kept secret inside of the strands in the light ocean.

Step 2: Black holes broke chunks away from the light ocean, releasing light to its full light speed.

Step 3: A black hole's gravity pulled light rays inwards. That slowed light's speed and released (imported) time to the area from light's tool box.

Step 4: The spin into atoms brought light into a duration of eons of being below light speed.

Step 5: Everything inside of this area, where light moved below light speed, experienced time.

TT says, 'There's another possibility. It just might be black holes who inflict time onto everything entering their gravity horizon. Space doesn't have time, while the time feature is invoked within the direct area anywhere around a black hole.'

TT adds, 'If black holes truly own time, then there's a need to reconsider their importance beyond anything we may have ever imagined.'

Short summary of new TT concepts.

Age and Origin

- T Theory sees an ordered cosmos – not chaos.
- Cosmos began small a trillion years ago.
- Cosmos grew to its present gigantic size.
- Cosmic growth was via 15 billion year cycles.
- There have been 66 such intertwining cycles.
- Our present 67th cycle hosts 73 quintillion stars.
- The stars in our sky are just the latest rendition.
- Cosmos originated from an endless light ocean.
- This endless ocean was static frozen light.
- The ocean was (is) the light supplier for cosmos.

Light and Black Holes

- Light was (is) the material spun to make matter.
- Orbs, solar systems, galaxies, came from light.
- Black holes are light eaters of the light ocean.
- Black holes spin light into atoms of matter.
- Black holes spin spheroids around themselves.
- A black hole exists inside of every spheroid.
- Black hole provides spin and gravity to its orb.
- Black hole's orb extends flat equatorial gravity.
- XL black hole inside a sun controls solar system.
- A larger black hole keeps more spheres in orbit.

Recycling and Supernovae

- Neither light nor a black hole can be destroyed.
- Light and black holes are the ultimate recyclers.
- Supernovae are the catalysts of the recycling.
- Solar systems recycle each 15 billion yrs, or so.
- Solar systems recycle (multiply) inside a galaxy.
- Black holes survive Supernova of a solar system.
- Survival black holes split in two after Supernova.

Building the next Solar System

- All the new naked black holes battle for light.
- Gravity rules the actions between black holes.
- Size matters: gives a black holes greater gravity.
- The black holes rebuild a bigger solar system.

Galaxies and Supermassive Black Holes

- Galaxies becomes hotels for solar systems.
- At hub of a galaxy is a supermassive black hole.
- Supermassives out-live the 15 billion year rule.
- Supermassives are up to 800 billion years old.

Space and Time

- Space results as light departures the light ocean.
- Space is a gigantic cavern inside light ocean.
- Light releases time when its speed slows.
- Time: duration light is slowed inside an atom.
- Perhaps black holes own the feature called time.

T Theory states there are changes we can make:

- Quit seeing cosmos in only primitive ways.
- Don't be fooled by what your eyes see.
- Always dig deeper for the answers.
- Think big, cosmos is complex beyond belief.
- Think like a designer as to how cosmos build.
- Apply strategy as to how cosmos was built.
- Envision fantastic science in the building
- Verify black holes to be the machines.
- Go out on a limb, that's where the fruit is.
- Quit trying to fit everything into a Big Bang.
- Replace Big Bang as paradigm model.
- Look at radically new ideas such as T Theory.
- Find the absolute proofs for Trillion Theory.

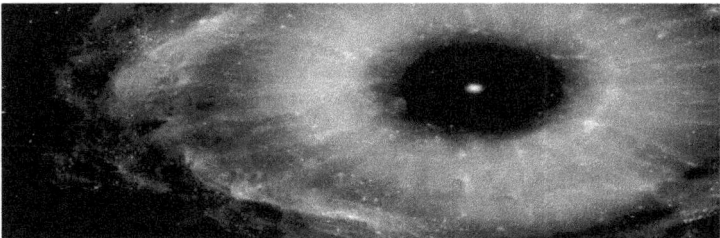

Show Me The Evidence. 5 TT Proofs.

T Theory Proofs

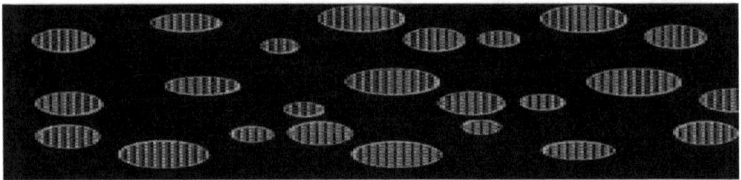

Proof 1 to authenticate T Theory (TT)(T).
Find black holes in a Supernova graveyard.

Doable. **TT says,** 'Powerful telescopes can peer yet deeper into the graveyard left behind after a Supernova (Obliteration) of a solar system. There will be many naked black holes who are survivors from the destroyed orbs of the old solar system.

If many black holes are found in the aftermath death of a solar system, that proves T. To find this proof, a telescope would need to peer through the overabundance of light pouring into the black holes. Johnny-On-The-Spot, for late to the party means missing the naked black holes battling to spin light into spheres for a new solar system.'

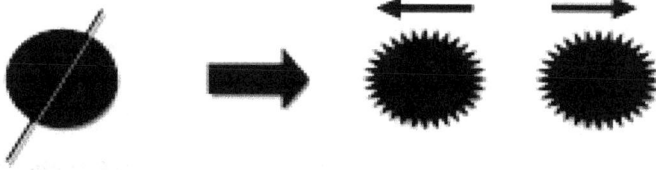

Proof 2 for T Theory (TT).

Axial spin direction varies independently with the various spheres (black holes) of a solar system.

If a nebular cloud supposedly spun coalescing to form our solar system (and TT says it didn't) then nebular spin should have given all the orbs in our solar system the same direction of spin on their axis, but it didn't. Nebular can't explain why spheres such as Venus rotate clockwise on their axis, while others like Earth spin counterclockwise.

This discrepancy is explained by TT: After the Supernova or Obliteration of an old sphere, the Replication split of the black hole which occupied that old sphere can arbitrarily apply either spin direction to the two newly formed black holes.

Proof would come by catching a XL black hole in its replication process following a Supernova. Or, proof comes from viewing any of the black holes in a Replication process following the Obliteration of their sphere and their destroyed solar system.

Proof 3 for T Theory (TT).

Axial tilts of solar system planets can differ.

Axial tilt explained. The axial tilt of Earth on its axis is 23.44 degrees. This means that when the Earth is on the left of its orbit around our sun, the north pole of Earth tilts towards the sun (meaning summer and more sun hours for northerners). A half year later, Earth on the right of its sun orbit, has its north pole tilted away from the sun (so northerners get more cold and fewer sun hours). Earth's tilt of 23.44 degrees is always the same, but different orbital positions cause the seasons.

Here's the rub. All the 8 planets orbiting our sun are in the same flat plane. But, Nebular Theory infers that every planet which was formed from the nebula should have the same axial tilt when riding in orbit. But, the 8 planets don't.

Neither Big Bang nor Nebular theory are able to explain this mystery. According to them, all of the spheroids should have gotten exact axial tilts.

But, they didn't. T Theory shows why.

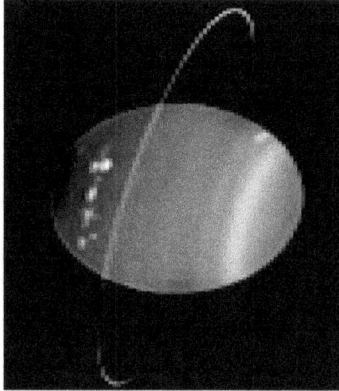

Planet Uranus actually lies on its side.

Astronomers have shown that the tilts of planets can differ; Nebular theory can't explain. Usually they explain that an off-kilter orb was struck by a passing orb, knocking it off balance. TT disagrees.

TT shows that each planet in our solar system operates independently under its own unique tilt. TT shows that each black hole that spun a planet around itself did so at a different axial tilt because each black hole searched for the best possible strategic angle for attracting light. That same tilt is still with the planet today as it orbits the sun.

We swing over to weirdo Uranus, spinning on its side. **TT says,** 'All black holes can independently determine their own axial tilt when they are in the midst of a battle for light to spin into matter. How Uranus got its axial tilt is a possible TT proof.'

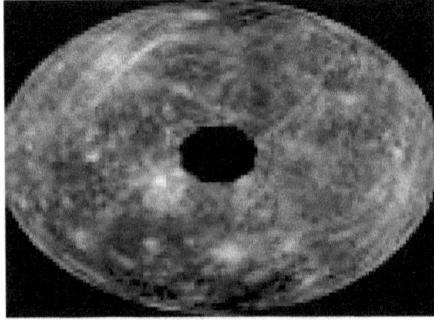

Proof 4 for T (Trillion) Theory (TT).

Find the black hole at the core of a spheroid.

T Theory says, 'If we could dig to the core of any moon, planet or sun, we'd encounter a black hole. This would be absolute proof of Trillion Theory.

However, the black holes we'd find at the core of those spheroids would appear different than any distant black holes which astronomers have perceived through their telescopes.

With spheres, it'd be difficult to differentiate the black hole inside of an orb from the rest of its outer body, since they blend as one. The black hole at the core, slowed by the matter it controls, gives the sphere its direction of spin on its axis, plus the gravity necessary to holding things on the surface while keeping smaller spheres in orbit.

It'd be a rush to travel to the inner depths of a sphere to discover its cloaked black hole.'

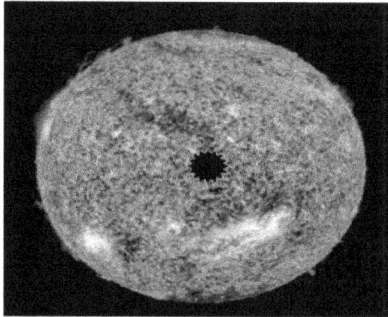

We might be surprised to find out just how small the black hole is that controls Earth. For instance, our sun has a diameter of 864,938 miles, yet the XL black hole inside is only 12 miles in diameter.

These cloaked black holes are mighty powerful for their size, having been built both for speed as naked black holes and also for power as cloaked black holes residing inside of spheroids.

The DISCOVERY of a BLACK HOLE in residence at the middle of our sun, which supplies gravity to hold our solar system in orbit around the sun, would be definitive proof of T Theory.

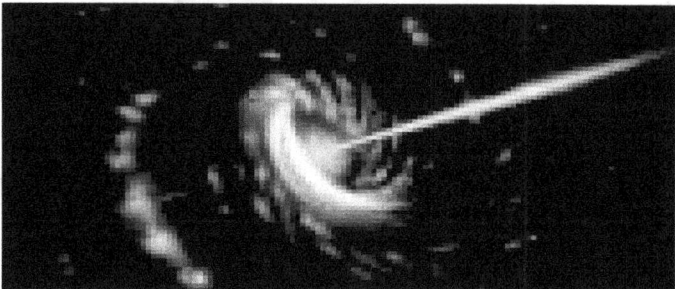

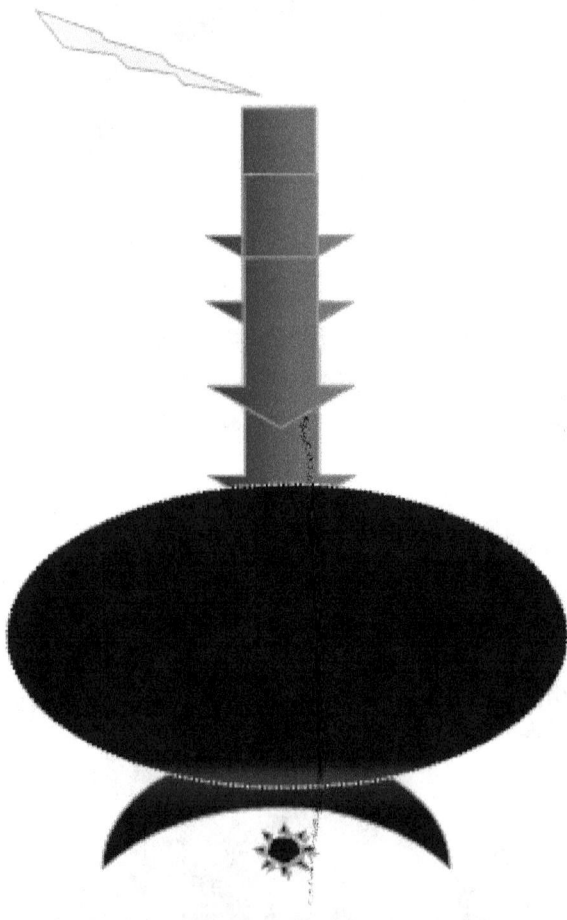

For Proof 5: The above diagram is a depiction of a Quantronix machine. T Theory says, 'The Quantronix ushers light down its long neck. The light is then coiled and spun at great speed in the wide belly. Finally, at the bottom, millions of atoms of a pure new element are produced.'

Proof 5 for T (Trillion) Theory (TT).
What if we take a black hole into the physics lab?
Spin light into matter in a physics laboratory.

 T Theory says, 'Spinning light into matter proves Trillion Theory beyond a shadow of doubt. In my futuristic sci-fi novel *The Trillionist,* the young lad invented a Quantronix machine which artificially did the task normally assigned to a spinning black hole. Quantronix was a large fat machine with a long giraffe-type neck sticking out the top of a domed building. This Quantronix then attracted light, spinning it into atoms of rare elements.

 The young lad succeeded at becoming the first alien to create matter from light. Perhaps, we humans can invent a Q-machine; become artisans creating new elements. Such a discovery would uncover numerous hidden cosmic secrets.

 A successful matter-making Quantronix would absolutely prove T Theory, showing the method that black holes deploy to form cosmic atoms.'

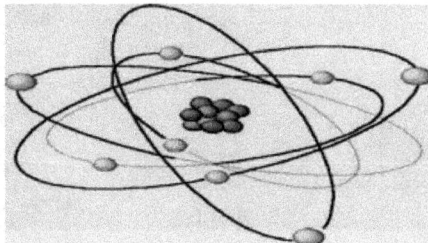

Astronomers and astrophysicists can assist with the finding of proofs for T (Trillion) Theory (TT).

Astronomers and astrophysicists will find many opportunities to prove TT. Proofs are out there that substantiate trillion year cosmos with black holes as the cosmic builders and operators. The time is ripe for a new cosmic paradigm. **TT proof is vital** for man's next giant step across space.

But for now, TT is mere conjecture, great fodder for discussion. Hopefully, in years to come, science will find the proofs for TT's advanced model. Proof of TT will leap-frog our understanding of cosmos and our universe. Then, Big Bang can be tossed out into the trash to join the incorrect 'Earth as center of the universe,' and the old 'flat Earth.'

TT says, 'Someday, man will climb high enough to peer into the Artisan's workshop to see how light was designed as the material, and black holes as the cosmic builders and operators.'

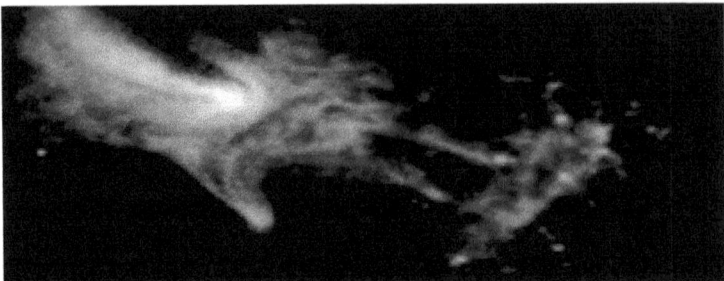

Biggest obstacles to proving T Theory.

Big Bang, although being a wrong theory, is very entrenched. Plus, astronomers struggle to find any rhyme or reason to black holes. They wrongly think that black holes are only formed after a sun goes Supernova, exploding and then imploding. It wasn't rationalized, until TT, that this black hole initially formed the star billions of years earlier.

Also, the incorrect 13.7 billion year age estimate for cosmos results when astronomers are fooled by what they see by focusing solely on the age of the stars in our present sky (can't see the cosmos for the stars), (can't see the forest for the trees).

Another obstacle: Man's time goes by in a blink. As actor George Clooney (of the movie *Gravity*) said about man's existence compared to cosmos: 'I'm very aware of just how brief life is.'

How to do it right

A Diane Grant quote, 'It's better to walk alone, then with a crowd going in a wrong direction.'

TT walks alone against millions in the crowd that follow a Big Bang in the wrong direction.

Do it right. Carl Sagan remarked, "Extraordinary claims require extraordinary evidence."

Let's say that T (Trillion) Theory is right, placing a black hole at the core of every spheroid. Why have astronomers gotten theories so wrong?

Even the very smartest minds can sometimes get conned into a paradigm such as Big Bang, where every shred of evidence has to fit a mold.

While Big Bangers still try to fit the discovery of black holes into their thinking, TT states, *'There wouldn't be a cosmos without black holes. Black holes are an absolute cosmic necessity.'*

T Theory shows how black holes operate like machines, organizing galaxies and building solar systems such as ours. The black holes have been doing cosmic recycling jobs for a trillion years.

'To prove T Theory, it's necessary to look for black holes in different ways: eating light when naked; in the graveyard of a Supernova of a solar system; inside of every mature planet or moon or sun; and at the helm of a galaxy.'

Defending T Theory from its critics won't be an easy task. For this moment, T Theory ideas may be a bit sci-fi, and extremely radical. However, on the flip side, the so-called experts who dispute T Theory, can't disprove it. Time will tell the tale.

It takes time to install a new paradigm such as TT which estimates cosmos to be a trillion years old and built by black holes. It was 150 years from when Copernicus first whispered that Earth wasn't the cosmic center until Sir Isaac Newton finally brought acceptance of this correct theory which placed our sun at the center of our solar system and Earth as simply one of the planets.

Here are some quite recent dramatic examples where man has had moments of being wrong:

- Western Union 1876, 'A telephone has too many shortcomings to be a means of communication.'
- Lord Kelvin 1899, 'A heavier than air machine which can fly, that's utterly impossible.'
- Charles Duell 1899, U.S. Patents, 'Everything that can be invented has been invented.'
- Robert Millikan, Nobel Physics 1923, 'There is no likelihood we can ever tap the power of an atom.'
- Thomas Watson 1943 chairman of IBM, 'There's a world market for maybe five computers.'
- Ken Olson 1977, founder Digital Equip, 'There's no reason anyone will want a computer at home.'

So, perhaps T Theory does have a chance.

Theory of Everything
ToE

TO BE THE CORRECT THEORY
For a theory to be the correct universe theory, it cannot ignore critical parts of our cosmos. A ToE must bring all the vitals under one umbrella.

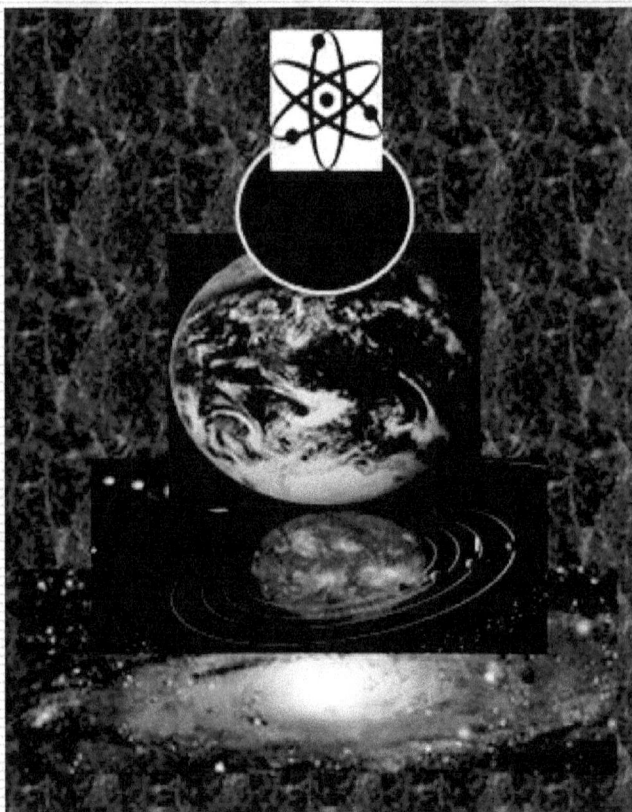

In the diagram, TT displays a correct ToE fitting each part of our cosmos inside a next larger part. TT says, 'Light and black holes are integral in connecting the dots from small to humongous.'

Atom: Subatomic particles fit into atoms where a black hole filled itself with light spun to matter.

A *Black Hole* resides inside of a planet such as Earth, which it controls with spin and gravity.

Planets orbit their sun inside of a solar system.

Solar Systems revolve around a supermassive black hole in control at a galaxy's hub.

An ancient *Galaxy* is but one of millions which travel throughout the vastness of space.

Space is an vast internal cavern created over the span of a trillion years by black holes eating away at the endless frozen ocean of light.

Light ocean is the endless frozen light material supplier existing on the outskirts of cosmos.

To conquer the final frontiers of space, we need a succinct Theory of Everything that encompasses the whole gamut, spectrum, panorama, domain, scope, breath and diversity of all that exists in our physical universe. right from the micro to macro.

Before TT dives in further, essentially recognize that a ToE is an extraordinary human feat. We've an insatiable thirst for excellence, so we best deal with this crucial subject from a factual perspective. It's unfortunate that 70 years was wasted invested in efforts trying to prove a ridiculous Big Bang.

But, it not too late for a lightning strike to wake us up to grasp the big picture. Extremely relevant is that T (Trillion) Theory properly connects the dots between atoms, spheres, solar systems, and galaxies, plus light, black holes, gravity, and space. Such TT connections make a proper ToE possible.

The Theory of Everything shown in T Theory puts forth a commonality from the quantum world of atoms all the way up the ladder to humongous galaxies. The common thread throughout is that *Light* spins forming matter and *Black Holes* are the engines of spin and gravity on every scale.

Here is the KEY

If we prove, as T Theory states, that black holes do spin light into matter, then everything in TT enters into a factual zone. Everything, right from atoms to galaxies hops onto a TT provable bandwagon.

A single black hole spinning light into atoms of matter would be a discovery, **but billions of black holes spinning light into matter makes a universe.**

TT says, 'While gargantuan beyond measure, the entire cosmos is simply explained by T Theory as orderly systematic arrangement of the subatomic, ranging in scale from atoms to immense galaxies. The architects of the spinning and arrangement of cosmic atoms is accomplished by black holes.'

Light spun by a black hole

TT says, 'Black holes are connectors explaining everything existing in cosmos. Black holes tie the tiny quantum world of atoms to spheres, to solar systems, and ultimately to galaxies. To thoroughly understand black holes is to understand them all.

The same principles are seen across the entire cosmos. On each and every scale, micro to macro, black holes are individualistically the shapers of light to matter and then on a group basis the organizers of all solar systems recycled inside of galaxies dating back a trillion historical years.

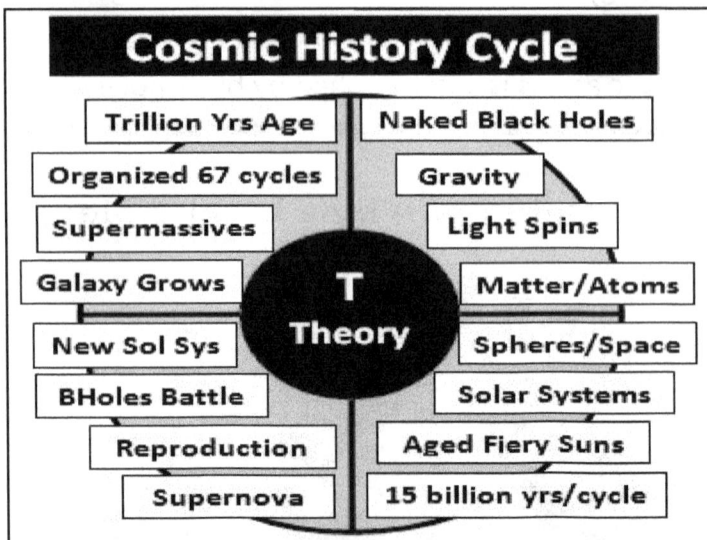

Cosmic History Cycle

Trillion Yrs Age	Naked Black Holes
Organized 67 cycles	Gravity
Supermassives	Light Spins
Galaxy Grows	**T Theory**	Matter/Atoms
New Sol Sys	Spheres/Space
BHoles Battle	Solar Systems
Reproduction	Aged Fiery Suns
Supernova	15 billion yrs/cycle

TT comes closer to a ToE each time we can link together more parts of TT's Cosmic History Cycle.

The Age Factor

TT's ToE places an age (youngest to oldest).

Organics: the shortest life cycle (100's of years).

Atoms, spun as matter, can be babyish on a new sphere or up to 15 billion years on an old sphere.

Solar systems, like suns and spheres comprising them, have an upper age limit of 15 billion years.

Galaxies are ancient when compared to spheres and solar systems residing in them. Resident solar systems die and recycle into new systems, within each 15 billion years, whereas the supermassive black hole at the hub of a galaxy has evolved and may be as ancient as 200-800 billion years old.

Black holes are indestructible, so some of the oldest are up to a trillion years, dating back to the origin of cosmos. Their age depends upon when they were introduced into our cosmos. New black holes (worker-bees) toil at the outer ocean of light. Whereas, a supermassive black hole is closer to a trillion years of age. TT count upwards of 73,000,000,000,000,000 quintillion black holes throughout our cosmos, of which a majority are cloaked inside of the sphere which they built.

Ocean of Light, dates back a trillion years.

Outside of the Ocean of Light, much older.

Suppose that T Theory is right. **TT says**, 'When it comes to explaining our universe we need to begin thinking like a top designer and as a clever strategist, for cosmos is specialized science.'

'Who holds the patents to the black holes which were deployed to build our entire cosmos?'

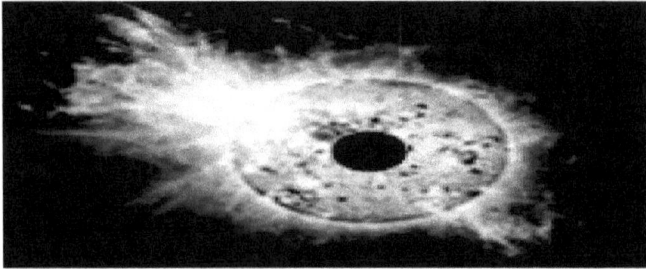

Cosmic
Design
Strategy

No other theorist has ever made as neo a theory as T Theory which approaches our cosmos from a designer's strategic perspective based upon the best which ultra-science could ever possibly offer.

T Theory alleges that much scientific genius was behind the design of our cosmos. This is seen by deployment of just one basic material - namely light, to hold a necessary vast array of properties to construct all the spheres of our cosmos. Then, equally great science is displayed in the design of black holes, enabling them to be the builders and the operators of cosmos. Black holes are so clever that they supply a method to recycle, grow, and perpetuate our cosmos from one generation to the next. Our cosmos is amazing engineering.

T Theory says, 'A magnificent grand design.'
There are inadequate superlatives to describe the magnificence of the science behind our cosmos. If it were a contest, our universe wins in a runaway.

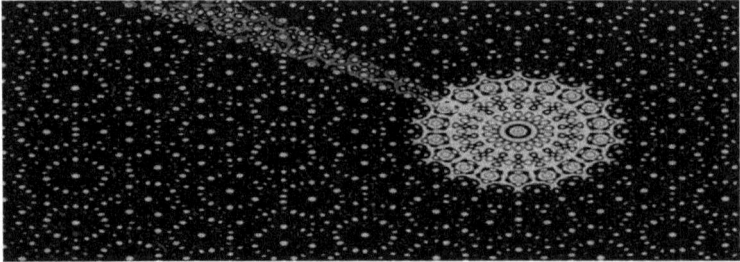

Your mission, should you choose to accept it

Suppose that YOU were given the challenge of creating a universe and the rules stated, *'Only one basic material and only one basic type of engine allowed.'* Stringent rules would make any type of grand cosmic design an extremely difficult task.

Both light and black holes are most worthy of being called beyond-super scientific feats. Just as amazing is an implementation of spin right from small micro atoms to gigantic macro galaxies.

In this sculpting, strategic design gave cosmos a hands-off operating basis with no manipulation. TT says, 'No has to run out to turn the crank.'

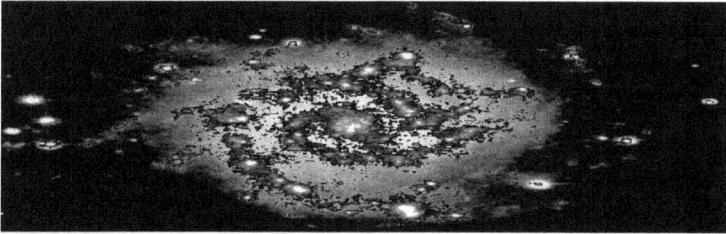

Such an amazing cosmos makes it easy for TT to emphasize its claim, 'Cosmos is no mere accident, nor pure happenchance, but rather a direct result which arose from a deployment of the pinnacle in scientific inventiveness, black holes being a prime example of this absolutely amazing science.'

TT predicts that in the near future the axiom and truism will be that black holes built our cosmos. It will be self-evident that Big Bang is at fault. For, here's the rub, black holes simply don't fit into the Big Bang. Whereas, a multifaceted T Theory does explain our complex universe to nearly a T.

T Theory further hopes that its new theories will become the acceptable cosmic paradigm. This will surely lead to chasing even bigger questions.

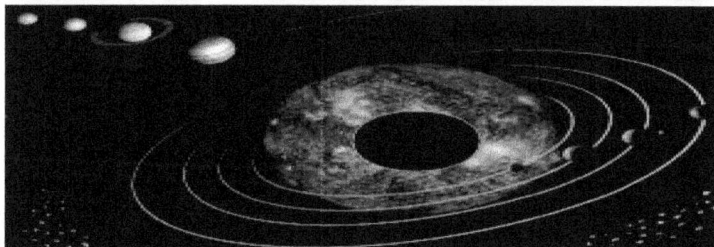

The following are some new ToE physical concepts uniquely designed for our cosmos.

The frozen Ocean of Light
It was in place a trillion years ago and still to this day supplies the building material for our cosmos

Space is a cavern in the Light Ocean
When black holes eat light away for the frozen ocean, space becomes the cavern left behind.

Light can spin to form atoms of matter
A yet to be proven property of light is that it can be spun by a black hole into an atom of matter.

Black holes build cosmic spheres
A black hole's structure makes it a sphere builder, spinning tons of light into matter around itself.

Laws of Gravity are set by black holes
Laws of Gravity between black holes prevail when they build and control solar systems and galaxies.

Black holes can grow larger and evolve
Over 100's of billions of years, dominant black holes adapted their structures which has allowed them to grow galactically supermassive.

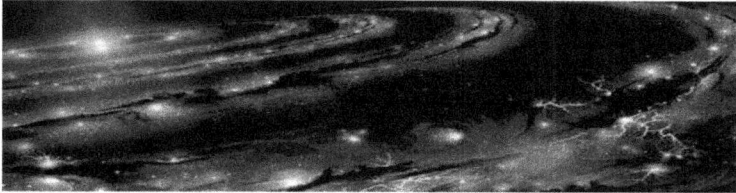

Does TT need to include who and why in a ToE?

For Big Bangers, the god thought or the word creation always gets them raising their fists. In dealing with this, TT prefers to take a scientific approach, leaving happenchance 99% out of the picture, while stating that whoever or whatever is responsible for our universe did so by utilizing a design and construction which is absolute genius.

Therefore, when TT speaks of the design of light and of manufactured black holes, then who built the cosmos, when did they build it, and why are we in it, may be more readily answerable once we solve all the design and engineering ins-and-outs.

TT suggests that who-why may transpire quicker once we solve our physical cosmos. Words such as creator, artisan, technician, corporation, or a computer hologram may simply be inadequate.

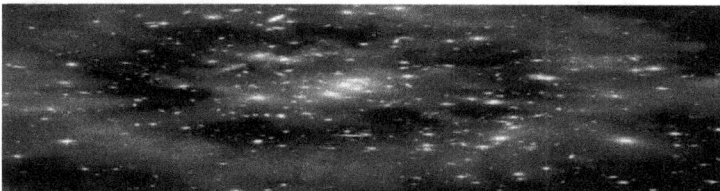

Organics and Spirits

T Theory readily admits that we, as a human race, may still be several Einstein type geniuses away from a Theory of Everything which bundles into one even more than just our physical cosmos.

However, TT can attempt to connect the cosmos with us living organics, plus the spirit world if it exists. Animals, plants, insects, et al, as organics seemingly have been built from a similar atom based structure as the physical cosmos. Plus, TT feels that a spirit world is not necessarily based outside of our cosmos as most religions tend to think. Remember, whomever did the design of our universe did so trying to keep all things recycling nicely inside of the cosmos; that might pertain to any spirit world as well, thinks TT.

Once again, TT feels that we, as organics and spirits, must look to black holes to gain a stronger understanding of cosmos and also our universe.

TT says, 'We should jump at the first chance to befriend a black hole. TT thinks they would open corridors for near instantaneous cosmic travel.

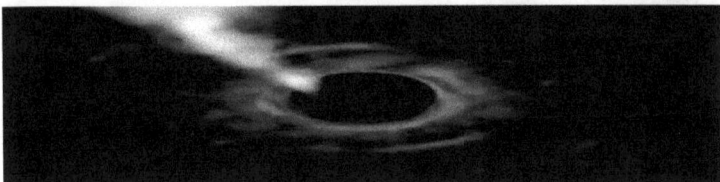

T Theory Says: Who Owns Our Universe
For TT, the huge question of the future will be, 'Who owns our universe? From outside or within?

From a physical perspective, TT has stated that we shouldn't tromp right past the obvious. If TT is correct that Black Holes built cosmos, and they are now presently in control of over 73 quintillion spheres, billions of solar systems, and millions of galaxies, then perhaps they are the quiet owners.

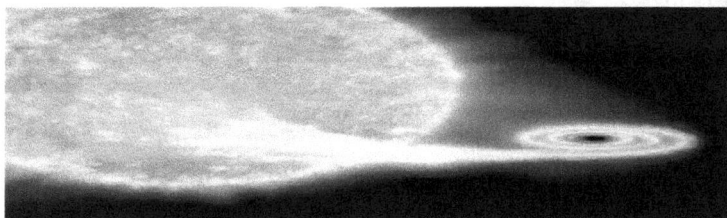

Black Holes Society
The members are the trillions of black holes across the cosmic expanse. The black holes battle with and against other black holes, erecting orbs, building solar systems within ancient galaxies. Just suppose that they do communicate with one another, while battling or co-operating. A type of society that takes very seriously their jobs of erecting an ever growing cosmos. If so, are they the cosmic owners, or is someone higher up.

TT says, 'We are on the verge of dethroning the Big Bang by discovering incredible hidden cosmic secrets about black holes. These discoveries will tell us: how black holes spin light into atoms; how black holes build cosmic spheres providing their spin and their gravity; how they arrange a solar system; how supermassive black holes marshal suns and solar systems inside ancient galaxies; how black holes recycle spheres and solar systems in galaxies every 15 billion years or so, during the trillion year history of our ever-growing cosmos.'

T Theory feels that its new theory really uncovers many of the traits and attributes which the owner of our universe possesses. When we understand cosmos, we then know what the cosmic owner was (is) working to accomplish.

TT makes an effort to place the main attributes belonging to a universe owner into a diagram

Attributes of the Universe Owner

[Scientific Genius] [Inventive] [Inspired]

[Goal Oriented] [Innovative] [Visionary]

[Imaginative] [Forward Thinker] [Artful]

[Super Designer] [Methodical] [Diverse]

[Resourceful] [Technician] [Architect]

[Intent for Detail] [Perfectionist] [Clever]

[Test over] [Trial Error] [Prototype Pilots]

[Patient] [Persistent] [Willful] [Adamant]

[Strategic] [Competitive] [Crafty] [Foxy]

[Complex] [Boundless] [Timeless] [Fair]

[Passionate] [Generous] [Humble] [Fair]

[Benevolent] [Powerful] [Wise]

[In Control] [Successful]

[Prolific] [Masterful]

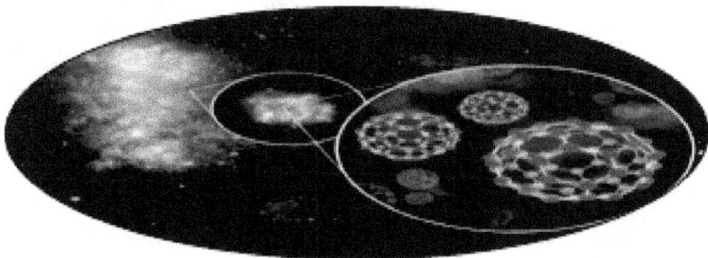

As this book nears its end, TT reminds the reader that the **main mission of T Theory** is the proposal of a new theory depicting how cosmos originated one trillion years ago, then grew to its present prodigious size via the major role played by black holes. TT is a game-changing view of cosmos.

What if our cosmos is a trillion years old?

What if black holes did indeed build cosmos?

If T Theory is true, there are huge implications:

• A trillion year old cosmos would be dramatic on both a physical and spiritual plane.

• If as T Theory suggests, every star has a solar system, this greatly ups the possibility of life on many Goldilocks planets (not too hot or too cold).

• If there has been human life on planets during the other 66 previous 15 billion year cycles, then life likely abounded during a trillion year history and achieved incredible technological levels.

• Some of these civilizations may have had the means to settle colonies outside their own solar system, to perpetuate their society and species.

• If reincarnation exists, then over a trillion-year universe, reincarnation may have a richer history than ever suspected. Read my novel *The Trillionist.*

The 2nd Goal of T Theory is to find Proof.

TT offers 5 possible ways to prove its theories using future scientific means. Such proof would make T Theory a new paradigm cosmic model.

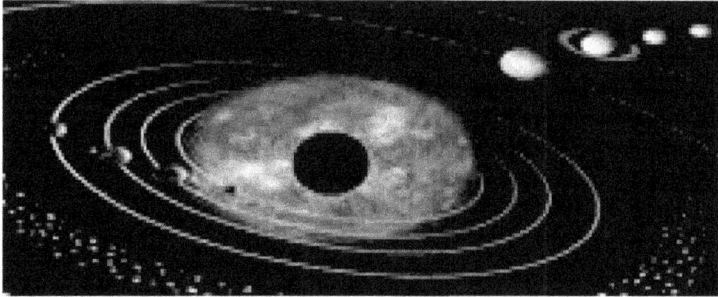

The 3rd Goal of T Theory: Say 'no' to Big Bang.

T Theory declares 'Big Bang never happened 13.7 billion years ago. It's an old theory, hanging on like a sacred cow waiting to be tipped over. In the near future, when TT's stated role of black holes is proven, then Big Bang ceases to be the paradigm.

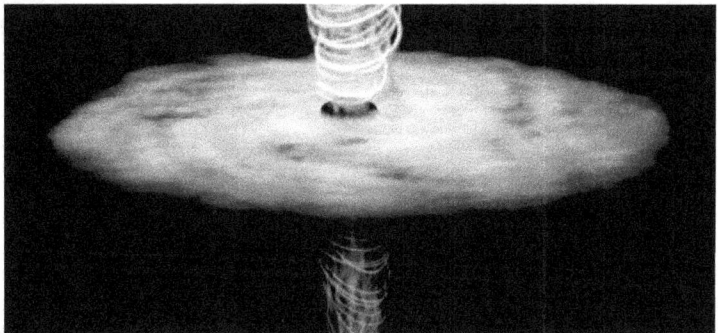

T Theory Says.

As founder (writer) of T Theory, Ed Lukowich is a *dreamer* type of person who has a propensity for seeing the *acme, summit, pinnacle, or zenith* in most things. Ed sees our cosmos and universe as a crowning achievement, worthy of winning top scientific achievement at any universe contest.

Yet, Ed also feels there is more, for every time he digs deeper, performing forensics on the cosmos, he encounters complexity taken to new levels. So this tells Ed that ours is not the only model which the universe owner has designed. Ed estimates that thousands of other universes exist, each with its own outstanding individualistic features, each far different than the universe we live in.

Thank you for reading this book. The ultimate hope is that you now think 'Trillion' and 'Black Holes,' as TT does. Or, at least some corridors have opened for you to new cosmic ideas.

No doubt, my new T (Trillion) Theory (TT) will be bashed and trashed by many. That's fine. My philosophy is that a person who opposes also helps a project. My passion outweighs my fears. For me, its a risk well worth taking.

My intent is to have no regret. For, T Theory is either right or wrong, accepted or rejected on its own merit. In the longer run, new scientific discoveries will provide the proofs. Such proofs may prove all of T Theory, or even some, which would be a major victory for a new paradigm.

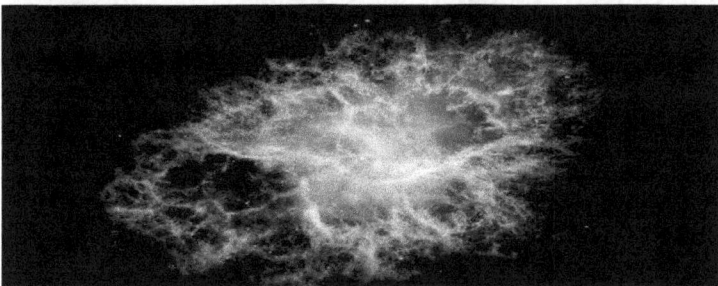

Carl Sagan said, 'It is better to grasp the universe just as it really is than to continually persist in delusion, however satisfying and reassuring.'

T Theory reiterates, 'As theory, TT is brand new. Many people see it for their first time and fear any reprisal for liking it. Thus, it will take time for this new paradigm model to become acceptable.'

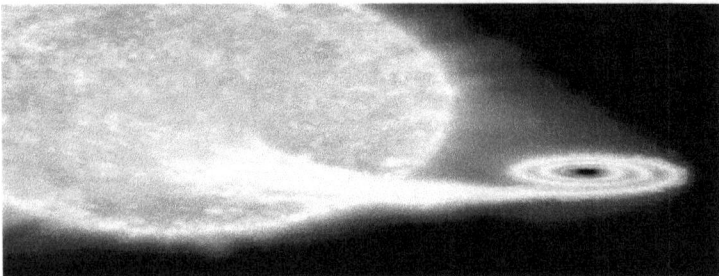

Author's Disclaimer:

The TT (Trillion) Theory theories in this book were developed solely by the author Ed Lukowich. These Trillion Theories are not based upon any other person's theories, regardless of any such claims.

No other person may claim to be the originator, founder, or author of any of these Trillion Theories.

So again, thanks for reading TT. You can be a T Theory supporter: website www.trillionist.com .

The future will bring many amazing cosmic discoveries. In 2018, NASA launches the James Webb Space Telescope as a new premiere space observatory to further study the formation of stars and planets and study birth and evolution of galaxies. Also, to hopefully discover alien life on other Goldilocks planets. Thus far, the Kepler program has found 1,000 exoplanets outside of our solar system. It'd be great to find evidence authenticating and proving T (Trillion Theory).

T Theory author Ed Lukowich (The Trillionist)

Books authored by Ed Lukowich

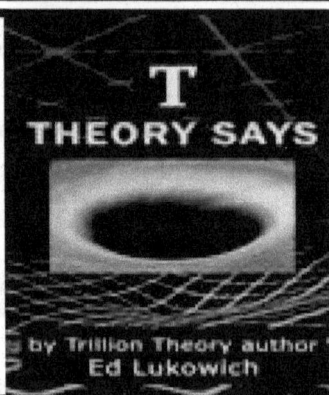

www.ingramcontent.com/pod-product-compliance
Lightning Source LLC
Chambersburg PA
CBHW060022210326
41520CB00009B/969